The Sizesaurus

The Sizesaurus

Stephen Strauss

KODANSHA INTERNATIONAL
New York • Tokyo • London

Kodansha America, Inc.
114 Fifth Avenue, New York, New York 10011, U.S.A.

Kodansha International Ltd.
17–14 Otowa 1-chome, Bunkyo-ku, Tokyo 112, Japan

Published in 1995 by Kodansha America, Inc.
by arrangement with Key Porter Books Limited, Toronto, Canada.

First published in 1995 in Canada by Key Porter Books Limited.

Library of Congress Cataloging-in-Publication Data

Strauss, Stephen, 1943–
The sizesaurus : from hectares to decibels to calories : a witty compendium of measurements / Stephen Strauss.
p. cm.
Includes bibliographical references and index.
ISBN 1-56836-110-6
1. Physical measurements—Handbooks, manuals, etc. 2. Size perception—Handbooks, manuals, etc. I. Title.
QC39.S844 1995 95-32176
530.8′1—dc20

Book design: Scott Richardson

Illustrated tables: Alfred Elicierto/Art Press
Sidebar illustrations: Brian Hughes

Printed in the United States of America

96 97 98 99 RRD/H 10 9 8 7 6 5 4 3 2

To my parents, Jack and Natalie Strauss,
whose support was immeasurable.

Contents

Preface to the U.S. Edition

Only the mad can measure Manhattan

My editor in New York has asked me to write a preface to the U.S. edition of *The Sizesaurus*. He thinks we may need to calm some American readers who'll see the metric enumeration in *The Sizesaurus* and become nervous at the unfamiliar units of measure. The first thing I want to say is that every metric measure has been converted into your more familiar units of inches, ounces, feet, etc. However, some of you may still feel uneasy, even with this double notation. Groups of people, indeed whole countries, become very attached to their familiar units of measure. For example, as an American, you know you have returned from foreign climes and currencies when you are transported once again to the land of the free and the home of the yard, mile, pound, and gallon. Familiar measures tell you, in ways that are difficult to express, that you are home again.

There is no avoiding what to many Americans will be the foreignness of metric in a book like this. However, the scientific and technical communities all over the world, including those in the United States, use metric, and if readers are going to put the measurement thesaurus sections to practical use, tables that include metric equivalents are absolutely necessary. And as disconcerted as some of you may be personally, dual measures will allow you to use *The Sizesaurus* in a world where—like it or not—metric is king.

My editor has also asked me to give an example of how comparative measures are arrived at. Tell the reader how big a New York City block is, he suggested. Fine, I replied, and took myself off to survey a street map of Manhattan. I quickly learned that it would have been wonderful if the founders of New York had laid out Gotham by some standard measure, because a simple glance at the layout of the city shows blocks in the form of pie shapes, or right triangles, or rectangles, or even weirder things. Times Square should more appropriately be called Times Siamese Trapezoid. I even discovered there is little agreement on the total area of Manhattan Island. One encyclopedia described it as 23 square miles or 57 square kilometers and another as 22.7 square miles and 58 square kilometers. My editor pointed out that these figures are impossibly contradictory. Could new landfill settling in the harbors explain the ostensible difference in area, he wondered. But I suspected a simpler truth that often turned up in my research: Reference books sometimes get their facts wrong. I then wrote to my editor that a Manhattan block is simply beyond measuring and slyly insinuated that if, as Thomas Wolfe told us, only the dead know Brooklyn, then maybe only the mad try to measure Manhattan.

"But," my editor swiftly replied, "most New Yorkers know there are twenty Manhattan blocks to a mile." But that didn't help me because when I went back to my map I quickly saw that blocks on which traffic flows east-west are clearly longer than blocks that run north-south. He admitted the difference, but responded that most New Yorkers understand that the twenty-blocks-to-a-mile rule refers only to north-south streets. What about crooked Broadway, I counterpunched electronically. (Most of our point-counterpoint was done via e-mail.) "Except for Broadway," he replied, which New Yorkers know was once a stream bed and a cow path and not a real highway. But, measure freak that I have become, I was not satisfied with this. "What about the crabbed and cramped streets of Wall Street, Chinatown, and Greenwich Village?", I asked. They clearly don't fit the north-south model, and were probably laid out on some old Dutch measurement. "Well," my editor replied, "most New Yorkers would know not to consider these odd blocks."

At this point, I had a moment of epiphany! Our exchanges were not a series of comic cross-cultural miscommunications, but an expression of this enterprise's most basic nature. In a thesaurus of comparative measures you want to use as references common objects that people can readily relate to. But when you search for some common standard, things get very, very relative. The typical length of a Manhattan block turns into a distance that only a savvy

New Yorker can understand and a unit of measure that has to have eight qualifiers to make any sense to the rest of the world.

So, please bear with the oddities of our relative measuring, gentle reader. In *The Sizesaurus* nothing is going to be quite as exact as some of you might like. Humbly speaking, I may even have made some errors. If you let us know about any that you find, we will correct them in subsequent editions. However, by creating useful comparative measures we may arrive at a system with which the world can be better understood and explained by chemists, engineers, cooks, carpenters, scientists, and just plain folks who want to convey, nonscientifically, how heavy, how tall, how hot, how loud, and how small things really are.

Stephen Strauss
Toronto
May 1995

Ack-nowledg-ments

Who to thank? The entire physics department of the University of Toronto. All the people who wrote all the books I used as references. The libraries of the University of Toronto, the Massachusetts Institute of Technology, the Library of Congress, and metropolitan Toronto.

On a more personal level, I have worked in the shadow of a master measurer. Joe Reid, my neighbor and metric conscience, went through a version of the manuscript and made numerous wise suggestions and corrections. No, I mask the truth. He caught many errors and saved me more heartache than I can think of. And he did it all, as would any good editor, with encouragement and not scorn. In the land of measurement Joe is, as T. S. Eliot said of Ezra Pound, *il miglior fabbro*. Hammond Dugan, a retired physics professor at York University, reviewed the finished manuscript and again saw things which had slipped by me (especially in the morass which is electricity and magnetism). I must also thank Cindy Krysac and Dan Falk who did research and measure translations for me. They too made interesting and useful comments.

Then, there are the many people on the Internet who answered questions and pointed out fallacies in my thinking. Particular thanks should go to André Roberge of Laurentian University in Sunbury, who put a face on the immensity of the Big Bang. Richard Stead, a seismologist in Virginia, helped me understand the nature of earth-splitting earthquakes. Thanks to York University, and in particular Judy Libman, for allowing me on their Internet. Numbers of people looked at and commented on individual chapters. David Wardle of the Atmospheric Environment Service and Wayne Evans of Trent University improved my chapter on the UV index, even when both of them disagreed with part of my take on the matter. John Adams of the Geological Survey of Canada read through my chapter on the Richter scale and—this is a theme—improved it. Harvey Levenstein, a McMaster University historian of eating, reviewed my chapter on weighing and measuring in cooking. And as great a contribution must be attributed to Una Abrahamson, whose wonderful collection of old cookbooks increased my knowledge exponentially. Not only did she have many books, but like all great collectors there was immense taste in those chosen. While on the subject of the cooking chapter, a doff of the hat should go both to Stevie Cameron and Pat Moffat, who so strongly encouraged me to write in this area. What I wrote was different from what they had originally suggested, but I trust we all learned something in the process.

I must also thank my editor at Key Porter, Renée Dykeman, not only for her many useful comments, but for the good cheer with which she dealt with the insecurities of a first-time author. Beverley Endersby showed me how careful copy editors can be, and the good questions they can ask. I also owe a debt to the *Globe and Mail*, which not only has allowed me to spend so many years learning about interesting things in science, but gave me six months off to work on the manuscript. I thank my family for putting up with my crankiness through what must have seemed to them a never-ending authorship. And lastly, but not leastly, Science Culture Canada helped make this book possible with a generous grant. No, again, I am too retiring. The book would not have been written without it.

Part I

Essays on assays: or, how to make a ruler out of almost anything

Intro-duction

Everything should be as simple as possible — but not simpler.
— Albert Einstein

Once, while still quite green in newspapering, I did something which many of you can probably relate to: I lied to get a job. An editor asked me whether I had any experience in layout — the term to denote design of the page — and I said yes, not a lot, but yes, certainly. A crash course over the weekend with someone who, as I saw later, was barely a graduate of the fingerpainting school of newspaper design got me in the door. Layout and I turned out to be an awful marriage — I once produced a page containing a photo of the Empire State Building and managed to crop off the distinctive spire in such a way as to make it look like Anything But The Empire State Building. However ugly my efforts, the experience turned out to be mind-expanding, because my struggles with design forced me for the first time to look at a page as something other than a corral for words. And one early summer night, as I ambled home with my fingerpainter friend, I had a kind of epiphany of layout.

Suddenly it was clear that the world had a coherent order. Signs and streets and the sewers beneath them had all been laid out by someone. The buildings, the cars, the clothes fitted over our bodies — all bespoke the language of design. And, to push it a bit, nature, with oaks and poison ivy and rabbit warrens and bogs, was merely the layout of things over which humans had no control — not to mention the solar system, the galaxy, the cosmos. In a burble I expressed my new awareness, my understanding of the majesty and the awesomeness of spatial relationships. My friend listened for a minute or two and then said, more sadly than sneeringly: "Yes, of course you're right, but it's one of those things which are so true that nobody gives a damn about it."

That remark removed me from my layout epiphany, but left me forever wondering how it could be that the most common truths were the least appreciated. It is a thought that underlies the seemingly diverse essays in this book. Coming intellectually, as I do, from outside of science, I have always been struck that the thing which most characterizes the scientific method is a fetish for weighing and measuring. I have come across more than fifty ways to measure color, a half-dozen systems for quantifying earthquakes based on Richter-type numbers, and enough psychological-behavior scales to sink the *QE2*. All scientists seem to understand, as Lord Kelvin so succinctly put it, that "if you can measure that of which you speak, and can express it by a number, you know something of your subject; but if you cannot measure it, your knowledge is meager and unsatisfactory."[1] What this has come to mean is that wisdom, at least as a means of understanding the natural world, is less important than an accurate scale. And as we have entered into the world of science and its handmaiden, technology, over the last four centuries, we have had to adapt to what is a most unnatural situation: both the microcosm and the macrocosm are being described with a precision greater than what our body's fleshy measuring rods can subjectively validate. If our ancestors' noses told them meat was rotten, they gagged and flung it to the dogs, but what does anyone's nose, tongue, or stomach now record when potential toxins come in parts per trillion? And because we don't understand the how and the what of measurement, the numbers it produces are accorded an iconic status. If an earthquake registers 7.2 on the Richter scale, then who are we to question; surely 7.2 is a number ordained by God or nature or some godlike someone. And just as surely, those of us who know nothing of its genesis or derivation must simply genuflect before such a cryptic quantification, much like Hollywood-extra natives worshipping the great god Kong. But such faith is groundless, simply because whenever we weigh and measure anything the process is more or less distorted. All numbers merely reflect the accuracy of existing technologies, but more than that, they are always embedded in a snarl of intellectual presuppositions. The reason the metric system increases by powers of ten has nothing to do with reality and everything to do with the absolutist madness of the French Revolution. There is no objective scale anywhere because all scales have objectives.

The inherent subjectivity of measure objectivity is *the* theme which transcends every individual topic in the essays that follow. It is a theme which carries with it a cautionary truth for all nonscientists who try to navigate the so-measured and therefore so-alien seas of the pervasive scientific culture. It is transparently true that all any layperson needs to know about most technologies is where the on/off switch is located, and the telephone number of a good repairperson. But measurements are different. If you don't understand their intrinsic subjectivity, their idiosyncrasies and weaknesses, then you simply don't understand. You become a prisoner of numbers and of the measure-world's numerous experts. Therefore, my cynical friend's putdown of that long-ago vision of layout does not apply to any similar *instant enlightenment* about measurement. When we are regularly asked to change behaviors based on somebody's findings using somebody else's scales — think of every diet you've ever tried — we have to give a damn. Anything less is measurement idol worship. As a kind of civic duty, like voting and paying taxes, we must challenge the experts and their measurements. We must say: Don't blithely tell me what you have found, Herr Professor, but tell me how you found it and what your measuring stick was. Because, if you don't, I will simply ignore you. I will trust my tongue, or my common sense, or just make up my own truths.

The conclusion that all of these essays reach is that it is the role of the scientist to answer readily and simply any layperson's queries about scales, and then to convey the ultimate truth: nothing is more absolute than the fact that nothing is absolute.

Speed kills, but does sound?

It is one of the most common and exasperated parental cries in the modern world: "Turn down that awful music before you make yourself deaf." It is hard to believe that it is only roughly forty years since the music became so loud, or since teenagers wanted to hear it at industrial strength.

The level at which sound becomes injurious has been quantified into a general world standard — namely, prolonged exposure to sound above 150 decibels (dB) can cause permanent hearing loss. Below that, there is a range of sound exposures that are benign for certain periods of time, but that's not what interests me. What I want to be able to say to my children is: "Turn down that noise. It's going to kill you, and maybe me." I want to say it because I want them to understand that it is not the words or music I object to, but rather a frightening physical force.

It turns out that there is some rough quantification of where death by noise might occur. But the best-researched numbers have nothing to do with the world of teenagers listening to heavy-metal bards and blowing out their ear haircells before they fry their brains. A whole series of studies have looked at the effect of ultrasound on humans. Ultrasound is effectively those sound waves which exist beyond the 20,000-hertz level that marks the high point of human hearing potential. Tests were undertaken in the 1950s and 1960s by people worried that ultrasonic waves in high-speed dental drills, jet plane engines, and cleaning machines were injurious to people standing near them. Animal studies showed that, when guinea pigs and rats were exposed to such sound at 152 decibels, they died. But what was ghoulish about the sound/death phenomenon was how freakishly it occurred. As the ultrasonic waves were absorbed into the body, they heated up the animals until Mr. Rat and Ms. Guinea Pig were literally roasted to death by sound energy. Humans standing near the animals, however, seemed unaffected. So, Horace Parrack, of the Wright-Patterson Air Force

Base in Ohio, did some calculations, taking into consideration humans' ability to dissipate heat — unlike rats and guinea pigs, we aren't encased in a shell of fur. He computed that, if you were exposed to 187 decibels of ultrasonic sound for 5 minutes, it would kill you. You could stand 185 decibels for 10 minutes, 183 for 20 minutes, and 181 for 40 minutes.[1]

He also computed that, if you stood 30 m (100 ft) from a jet engine, the ultrasound would have to be 200,000 to 500,000 times stronger to turn you into a roast. Even mechanics standing 7.6 m (25 ft) from jet plane engines would be exposed to levels 10,000 times weaker than those which could possibly kill them at any time exposure.

Of course, there are other ways that noise can affect you beyond what amounts to death by sonic roasting. The normally reliable *Guinness Book of Records* for 1994 suggests that someone, somewhere, computed that a kind of lethal overpressure wave occurs above 192 dB. This makes sense. All sounds are pressure waves and, as such, can shake the body back and forth until our internal organs are jellified. Think of it as being smashed thousands of times per second by waves as you stand at the seashore. This damage to internal organs is what happens when the shock waves associated with nuclear bomb blasts quash into people.

While *Guinness* had a precise-sounding number, unfortunately it's not one sound researchers are familiar with. Dan Johnson, a scientist at the E G & G Company in Albuquerque, New Mexico, briefly experienced sounds that registered at 196 dB while wearing special ear protectors. "And I'm still cooking and talking," he jokes.[2] At 192 dB, according to his calculations, a sound wave has a pressure of about 0.8 of an atmosphere, or 8/10 the air pressure we normally experience at sea level. The lethal atomic blast waves weigh in at four or more times higher than that. Neither Johnson nor anyone else can point to a single study which correlates a given decibel level of audible sound — as opposed to a shock wave — with a human's being shaken to death.

There must be a moral to all of this. The easiest thing to do is to clap one's hands and cry that even that secular bible of our times, *The Guinness Book of Records*, is fallible. (When questioned where their figure came from they ultimately admitted to me that the folder where the information was supposed to be was empty. And — in what may well be the high point of my life — that my research had led to the removal of the entry.) But beyond that it is clear that we have stumbled upon a truth about lethal measurements and humans: we aren't going to learn the upper ends of some scales unless we kill people to discover them, and, Nazi science excepted, the world has been loath to go that route. In other words, the answer to the question of how loud a noise would have to be to kill us is: too morally loud for science to calibrate.

Big Mac measures the universe

It used to be easy to measure things. You would pick a king, almost any king, see how big his foot was, and call that a foot. Then you would measure the length from the end of his nose to the tip of his forefinger and declare that a yard.

But today is different — kings are in short supply, and even if we had them we would want measures with more universal natures, even perhaps the often weird measure equivalents you find in Part II of this book. From the first moment I conceived of *The Sizesaurus*, the question I have been struggling with is what in today's world is both common and universal and uniform? What transcends culture, class, language, age, sex, *and* religion? What produces man-made things in shapes and sizes which are so ubiquitous that we almost cannot differentiate them from the natural world?

And then in a flash, or rather in a kind of half stomach growl, half burp, the answer came to me. McDonald's, the world's biggest restaurant chain. Its claim to fame is not only that it serves food fast but that every Big Mac, every fry, every milkshake is a repetition of every other one, from Beijing to Moscow, to Boise, to Brazil. Others may decry the elimination of variety in McDonald's endless efforts to be the same, but *The Sizesaurus* embraces it. We measure out the world in Things McDonald's and then come back for seconds.

Let us start with that star of double patty, pickle, and sesame seed bun — the Big Mac. It weighs 215 g (7.6 oz). It is more or less 5 cm (about 2 in) high. This means that a person 183 cm (6 ft) tall is the same height as a 36 Big Mac stack, and the world, which weighs in at 5.98 x 10^{24} kg (1.32 x 10^{25} lbs), would be balanced on a cosmic scale by 27,800,000,000,000,000,000,000 Big Macs. Yummy. Volume-wise, BMs are roughly 410 cm^3 (25 in^3), which means that 6,100 of them would fill up an average shower stall, and 54,000,000,000,000,000,000 would be roughly equivalent to the volume of the Moon. A Moon made out of Big Macs — it's something to stir the imagination of the most measurephobic.

The cooked hamburger patty is roughly 7.6 cm (3 in) in diameter. Thus, if a road 120 patties wide were built between New York and San Francisco, 78,000,000 all-beef paving patties would be required. If you wanted to pave the same road with the sugar packets that come in the restaurant's 24-cm^2 (3.7-in^2) -sized containers, you would need 15,500,000,000 of them. To make a square kilometer, you would need 420,000,000 sugar containers or about 55,000,000 Big Mac patties. A square mile is equivalent to 2,800,000,000 sugar packets or 140,000,000,000 Big Mac patties. A square meter is equal to 800 of the little McDonald's salt and pepper packets (each about 12.5 cm^2, or 1.9 in^2); on a larger scale, it would take 6.6×10^{16}, or 66,000,000,000,000,000, of them to cover the surface of the Atlantic Ocean.

(In all frankness I must make a confession here. There can be a slight difference from country to country in such things as condiment containers, napkin sizes, straw widths, etc., so what I am giving you is McDonald's Canada Ltd. metric. However, since we are in the realm of the more-or-less measure, the McDonald's public-relations people have assured me that there are probably no dramatic changes between countries.)

So that brings us to those things which come in volumes. A McDonald's straw will hold 7.7 ml, or just over one-and-a-half teaspoons, of whatever liquid you are slurping through them. This means that it would take 17,000 strawfuls of water to fill up a 130-liter (34-gallon) bathtub, and 130,000,000 to fill a typical two-story house.

And let us not forget those McDonald's drinks which come in child-size (250 ml), small (345 ml), medium (495 ml), and large (730 ml) containers. You could fill the world's largest oil tanker with 1,020,000,000 child's-size, or 739,000,000 small, or 515,000,000 medium, or 349,000,000 large drinks.

If you don't like any of those measures, then construct your own, using the 37.5-ml (6¾-tsp) plastic container of barbecue sauce or the 17.5-ml (3½-tsp) container of honey. Both are used to flavor Chicken McNuggets, which themselves come in a 191-cm^3 (about 11.7 in^3) box.

Lastly, for the diet-conscious, keep in mind that the energy consumed by eating one BM would allow a 73-kg (160-lb) man to play golf for just under two hours, or to watch TV for nine hours.

I place no value judgment on this exercise in McMensuration, except to say that, in an age in which the fast and uniform meal is mankind's taste preference, one additional and completely subversive truth may emerge. If we are what we eat, then everything supposedly subjective about us, including our thoughts, dreams, loves, wars, and weeping in the night, might have some kind of standardized McDonald's equivalent. I don't know the scale, but I do know this measure's name: a Big McSoul.

How big is a big bang?

Big things — really, really big things — are hard to put into perspective because they fall off any sensible scale. Take the universe. I used to think it was as big as infinity, but astronomers have recently calculated that it really is about 8 to 12 billion light-years across and, consequently, 8 billion or 12 billion years old. And of course anything that scores in the range of eight billion to twelve billion is hardly worth a mention because even a country as middling important as Canada has a yearly deficit bigger than that.

Which brings me to that huge explosive event, the Big Bang, which created the universe. On page 132 of the classic first-year physics textbook *General Physics,*[1] there is a chart of approximate energy releases for a variety of natural phenomena. And at the top is the Big Bang. It is described as being in the order of 10^{68} joules. Now there are several ways of understanding the scope of this number. Expressed less scientifically, the number becomes 100,000,000,000,000,000,000,000,000,000,000,000,000,000,000,000,000,000,000,000,000,000,000,000,000,000 joules. To put that into perspective, the same book gives the amount of energy used annually in the United States as 8,000,000,000,000,000,000 joules. Which actually doesn't put it into much perspective at all if you don't have a feeling for what one joule might be. Try considering that the daily food intake of a North American adult is about 10,000,000 joules or, in dieter's measure, about 2,390 Calories. Still confused? How about pondering the fact that the amount of kinetic energy which it takes to hit a home run is 1,000 joules and that to turn this page will expend 0.01 joule. Still not happy? Well, I don't blame you, because something in the back of your mind may be asking how we could ever measure how much energy the Big Bang produced. And the answer is, as André Roberge, of the Department of Physics at Laurentian University in Sudbury, Ontario, put it to me: "It's *not* possible to derive such a number without making some assumptions which nobody can verify." Indeed, being unable to find any generally agreed-upon way of arriving at this figure, I contacted Professor Morton Sternheim, one of the co-authors of *General Physics*, who confessed that he had not kept a reference to where the figure had come from and couldn't really provide any deep insight as to its origin. It is what is known in the physics-problem game as a "Fermi problem," named after Nobel laureate Enrico Fermi, who delighted in forcing his students to figure out ways of guesstimating answers for questions involving huge or tiny numbers.

Professor Roberge made a stab at arriving at a Fermiesque Big Bang number, admitting all the way that there was a lot of handwaving involved. If we believe that the observable universe contains about 10^{78} protons and neutrons, and each proton or neutron has a tiny bit of energy in it, we can construct an equation. The energy in protons and neutrons when they are not moving is about 10^{9} electronvolts, and 1 electronvolt is equal to 1.602×10^{-19} joules. Therefore, if we multiply $10^{78} \times 10^{9} \times 1.602 \times 10^{-19}$ we get about 10^{68} joules.

Fine, except that others have computed the proton/neutron figure at 10^{76} or 10^{80}. It doesn't seem like a significant difference, but anything which is a hundred times bigger or a hundred times smaller than an enormously large number is a number almost beyond comprehension. A hundred of the 10^{68} Big Bangs could exist inside a 10^{70} joule universe.

Wow, or should that be Wow^{70}? There are several worms in the physics woodwork, however. The observed universe is not necessarily "the" universe, and indeed, astrophysicists have continually speculated that upward of 90 percent of the universe exists in an unseen form called dark matter. But worse, Professor Roberge points out that the equation ignores gravitational energy in the universe: "It is conceivable that once that quantity is defined in a consistent fashion, it will be found that it is of the order of NEGATIVE 10^{68} joules, so that the total energy released by the Big Bang would be zero!!! In other words, the Big Bang would be nothing but the conversion of zero energy in the form of no spacetime and no particles into zero energy in the form of spacetime plus particles."

I am not going to pretend I understand half of this, but such a Fermi number does lend itself to a Zen-like joke. How big was the Big Bang? you ask. So big, replies the physicist, that all the energy in it amounted to zero. And how big is zero? you then respond. As big as the Big Bang, he answers, as big as the universe.

How hot is Hell, anyway?

The English language treats Hell as a place where measure anomalies abound. Somehow Hades can be both "hot as Hell" and "as cold as Hell." This contradiction has led some disgruntled Canadians to opine that Ottawa, their hot/cold capital, *is* Hell, or at least Chicago north. But that is a discussion for another day. Our question is simpler: Is there a way of determining how hot Hell is, and how cold it would have to be for it to freeze over? It turns out that there is a venerable calculation for the climate of Hell. It is attributed to a mysterious Mr. Wensel of the U.S. National Bureau of Standards and was reputedly made more than a half-century ago.

He proposed that the searing statement in Revelation 21: 8, "But the fearful, and unbelieving . . . shall have their part in the lake which burneth with fire and brimstone," provided a temperature touchstone. In order for a lake of molten brimstone to exist, the temperature in Hell would have to be below 444.6°C (832.3°F). For, if it were not, goes this argument, the brimstone and vapor would have become a gas and not a lake.

However, Hell's relatively low temperature produces a divine anomaly. In Isaiah 30: 26, there is a depiction of Heaven: "Moreover the light of the moon shall be as the light of the sun, and the light of the sun shall be sevenfold, as the light of seven days." When all this heating multiplication is rolled into one and the Stefan-Boltzmann fourth-power law for radiation is applied, it turns out that the angels are floating about in an ether heated to 525°C (977°F). Ergo, by this calculation it may be hot in Hell, but Heaven is even hotter.

Alas for the simple minds who would like to blithely depress the pious with this paradox. It seems that every time someone gleefully trots out Mr. Wensel's numbers, some skeptic points out that an earthly assumption is at work — namely, that the pressure in Hell is the same as the pressure on the Earth's surface at sea level. And that surely is wrong, if for no other reason than Hades's placement in the area we euphemistically term "below." Under intense pressures, brimstone can stay liquid to 1,040°C (1,904°F), but it would take only about 4.2 atmospheres, which is a bit less than our veins can withstand without rupturing and less than 1⁄17 of the pressure a 50-kg (110-lb) woman would exert at the tips of her stiletto heels, to make Hell hotter than Heaven.

But soft, if we assume that there is enough pressure — 32 atmospheres — to reach the 1,040°C figure, then a series of back-of-the-envelope calculations indicates that the temperature would have to fall to about 116°C (about 241°F) — hotter than the temperature at which water boils at sea level — for the lake of sulfur to turn solid. That goes up a bit if we increase the pressure enormously — something like 800 atmospheres — but it appears that 116°C (241°F) is about as cold as any reasonable Hell would get before its sulfur lakes froze over.

Thus we might suggest that, the next time anyone complains about the muggy heat of an Ottawa or Chicago August, he or she should think again. However steamy the city gets, it is still much, much colder than Hell.

Toward a Santa-metric

J. Alfred Prufrock may have measured out his life in coffee spoons, but children, ah, children, demarcate their lives in Christmas presents. And because they do, it seems particularly appropriate to describe to them (and us) just how hard it is for Santa to deliver his load of toys.

I do this not to be a skeptic or a wisenheimer, but rather because Saint Nick's travails give a human face to the often overly abstract effects of physics on earthlings. I also do it because, like getting Christmas presents, a Santametric is fun.

The first part of this essay owes its inspirations and most of its numbers to acidic calculations the satirical review *Spy* did a few years ago. Since that time, the computations have become a source of delight to Santa-doubting physicists everywhere. The observations go as follows: There were, circa the 1991 census, roughly two billion children in the world. By children I mean people under the age of eighteen. Since Moslem, Buddhist, Jewish, Hindu, Sikh, and animist children likely would not be visited, we compute that about 378 million children will have to be given presents. Let us say 15 percent of the Christian kids were naughty and not nice, and thus deserved to be rewarded with only the back of their parents' hands. With a census average of about 3.5 children per household, there are still 91.8 million houses, apartments, and other dwellings on the Ho-Ho-Ho route.

To accomplish his mission, Santa has 31 hours of Christmas to work with. The added time comes as a benefit of time-zone changes and assumes that he travels east to west to take advantage of them. This means that he has to make 822.6 visits a second (excluding whatever time Santa and the reindeer need to attend to bodily functions). His visit schedule translates into 0.0012156577 of a second to do everything — sleigh-parking, chimney–jumping-down, cookie-and-milk–consuming, presents-laying, and then reversing gears and getting back to the sleigh and off to the next stop.

While these visits would not be distributed evenly around the globe, for the sake of the calculation we are going to assume they are. This means that Santa, sleigh, and reindeer would have to travel about 1.25 km (0.79 mi) per household. This would bring the total trip to about 120.8 million km (75.5 million mi).

To achieve his toy-giving goals, Santa's sleigh would have to move at 1,040 km (650 mi) a second. This is 3,000 times the speed of sound and considerably faster than the 24 km (15 mi) an hour which a nonflying reindeer can hit at top speed. Indeed, the fastest man-made vehicle, the *Ulysses* space probe, reaches only a relatively lethargic 44.4 km (27.4 mi) a second.

And then there is the sleigh payload problem. If the child gets one medium-sized Lego set weighing in at about 0.9 kg (about 2 lbs) the sleigh must weigh 321,300 tons, Santa excepted. (Clearly, even if he were grotesquely overweight from all those cookies, it is obvious that the Jolly One would not make much difference in the sleigh weight.) A conventional reindeer, whose shoulders are only about chest high on your average adult, can pull about 134 kg (300 lbs). Although there is no indication that reindeer of any sort can fly, if we assume that there is some undiscovered winged reindeer species that could pull ten times as much weight while flying, it would still take 214,200 of them to pull the sleigh, give or take one, depending on Santa's weight. This means that reindeer and presents alone would weigh more than four times the weight of the *QE2*, something more than 353,000 tons. It is fair to say that this flying present-delivery system has not been shielded with added tons of heat-absorbing tiles, as is done with space shuttles. So when they fly at 1,040 km (650 mi) a second, Santa's team is going to encounter enormous air resistance. Each member of the lead pair — let's call them Cinder and Ash — will absorb 14.3 quintillion joules of energy per second. This will, of course, cause them not only to turn as red as Rudolph's nose, but to burst into flame almost immediately. As they are vaporized it will then become the turn of the deer behind them. The entire team will turn to nothing in 0.00426 of a second.

Even if by some miracle this didn't happen, Santa himself would not survive. If he weighed 113 kg (250 lbs), he would be smashed back into his sled with roughly 19,000,000 newtons (4,300,000 lbs) of force.

Although a skeptic would conclude that Santa is either dead or nonexistent, a believer in human inventiveness might suggest that all of Santa's problems will soon be solved when the world is hooked up to fiber optics. Instead of flying, Santa could send computer games, digitized movies, and virtual reality Barbie dolls to all the good little boys and girls over juiced-up phone lines. When and if the so-called Information Highway is installed worldwide, according to my computations Santa would be able to move six billion bytes of information a second. With this power at his command, he could "visit" all the children's homes on Christmas and leave behind something like the popular Tetris computer game, which weighs in at 444,000 bytes. Claus would be able to deliver 13,513.5 games a second, 810,810.8 a minute, and 48,648,648.649 an hour. Thus, in a shade under 7 hours, 43 minutes, and 44 seconds, Santa could deliver all his goodies without ever having to experience the horror of atmospheric physics.

Of course, this scenario would put all the reindeer and elves, and even Santa himself, out of business, but that's the modern world for you. Technological change makes everything easier, except holding on to the job you were trained to do.

Cold enough to freeze the miscon-ceptions off a brass monkey

There is an odd and persistent slang measure of cold which may be unique to the English language. It says that, at some frosty point, it is "cold enough to freeze the balls off a brass monkey." Cool, I thought, when I began to write this book — it was the depths of what seemed an endless winter — finally, I will be able to quantify how cold brass-monkey ball loss must be. So, I (nonmetrically) created an imaginary monkey with gonads weighing about 113 g (4 oz) each, joined to the body over a space of about 6.45 cm^2 (1 square in). I then took my measures and cast them upon the computer discussion networks which many people are tied into these days.

The answers I got back made my query look like the work of a dummkopf. "Didn't I understand the source of the image?" people asked me. Rather than a real monkey, what was being discussed was a military object. Specifically, the monkey in question was a triangular-shaped thingamabob, something like the rack used to set up balls on a pool table. In the nineteenth century, it was used to hold cannonballs. When the temperature dropped quickly, the brass monkey contracted and the iron balls didn't, at least didn't very much. The sudden movement forced the cannonballs to pop out of their brass corset. Hence, the expression. One of my correspondents, Raymond Gilbert, of the University of Indiana, went so far as to calculate the parameters of the brass monkey, the cannonballs, and the temperature at which the popping state would occur. By his calculations, when a boat that had been sailing at 24°C (75°F) found itself in air temperatures of roughly –26°C (–15°F), the balls fell out. However, he mistakenly assumed that the iron balls were actually made of lead, and another correspondent pointed out that, according to the dimensions he gave them, his popping balls would weigh about a ton each, not to mention the fact that lead deforms easily.

All of this sent me to the library, and in particular to language reference sources, which I thought might lead me to a primary source. You know, a real measure of the size of a brass monkey. Eric Partridge, who is sort of the Webster of slang dictionaries, suggested that the phrase was of Australian derivation. That was something the Australians apparently were not aware of, since none of the half-dozen books of Australian slang mentioned it. Partridge went on to claim that no one thinks of the term's civil origins today. Everyone thinks that we are talking about monkeys made of brass, maybe of the hear-no-evil, see-no-evil, speak-no-evil variety. This is shown by the various humorous efforts which have been made to soften the sexual innuendo in the phrase. Consider "cold enough to freeze the brass buttons off a flunkey" and "cold enough to freeze the brass off a bald monkey." Robert Hendrickson, in *The Henry Holt Encyclopedia of Word and Phrase Origins* (1990), weighs in with the remark that "nobody has yet proved the existence of any rack called a 'monkey,' one made of brass or anything else." Indeed, the *Oxford English Dictionary* reserves "brass monkey" for a kind of cannon. Naval dictionaries were devoid of brass monkeys. Naval historians didn't even know the term referred to sailing ships. Desperate at this point, I tried the question out on the computer discussion group which reference librarians have developed to find answers to impossibly hard questions. They call it "Stumpers." Brass balls stumped them.

Suddenly, it dawned on me that perhaps I had been led astray by my so-certain computer sources. Where did you hear it from I asked, and got the same answer back from everyone. From somebody who had heard it from somebody else. Ah, I said to myself, a nineteenth-century, sea-going urban myth. So I decided to return to my original tack. I approached metallurgist Robert Messler of Rensselaer Polytechnic Institute in Troy, New York, and posed my brass monkey brass balls problem to him. He suggested it was a hard question to answer because brass is an alloy of copper and zinc, and its brittle fracture point changes from alloy to alloy. Commonly used commercial brass consists of less than 40 percent zinc, and that means the metal would never become brittle enough to break. "Never" in this case means even if the temperature fell to absolute zero, that is, –273.15°C (–459.7°F). However, if you raised the zinc content, the metal might become brittle at around –100°C (roughly –150°F). This would seem cold enough to freeze anything off everything, but if you are looking for a measure more in the realm of human experience, one of my computer correspondents suggested that the temperature of –40 officially be designated as "cold enough to freeze a witch's tit." The appellation arises because it is at –40 that the Fahrenheit and Celsius scales converge.

Whoa, I thought, another English-language cold metaphor. I wonder where it comes from? And then I thought again, and went on to the next chapter.

The metric-saurus

A sharp-eyed observer will note on the frame of my basement door a set of historical markings. More precisely, there is a penciled line, wiggly and inexact, under which is written "February 21, 1992." Over it is another line, separated from the first by roughly the size of the nail of my baby finger. It says "August 10, 1992." The distance between the two represents a growth spurt of not quite six months, which took place during my daughter's twelfth year.

Like many parents, I have traditionally looked upon this handmade record of the physical changes in my daughter from a warmly biological perspective. Nature and nurture *are* working; here is the indisputable evidence that Anna is becoming a full-sized and ripe human being. But of late I have taken to staring at the size difference, and I realize I am seeing something else there. It is a measure unto itself. One could, if so inclined, make Anna the measure of all things — or, at least of all lengths. We could take the distance between February and August and spin out of these kilo-Annas and centiAnnas and milliAnnas and all the other players of the ten-times-ten-times-ten chain of enumeration which the world calls metric. But unless something quite unbelievable happens, we won't.

We won't, because humankind, even with the United States recalcitrant to the last, is nearing the end of what has been a quite incredible journey out of measurement anarchy. It is not quite fair to say that before the standardized system which is now officially entitled Le Système international d'unités (SI) came into being that every man and woman was the measure of whatever was being measured, but it is not far off. Examples abound. In the eighteenth century, in Lincolnshire, England, a "brescia" was the amount of turf a man with a spade could dig between May 1 and August 1. In Wales, a "leap" was 2.057 m (6 ft 9 in), or the normal jump of a working man. The French had a *houpee*, or the distance over which a stationary man could hear the shout, the "houp," of another man.

The corollary to individuals being the measure of all things is that everyone and everything had its different measure. Eighteenth-century France, the birthplace of metric, was a weights-and-measures swamp. There were thirteen different feet, ranging from 27 cm (10.6 in) to 34 cm (13.4 in); twelve different pounds, ranging from 344.13 g (0.758 of a pound) to 518.88 g (1.143 lbs); and twenty-three different bushels, whose volume ranged from 2.54 liters (0.671 of a gallon) to 102.3 liters (27.2 gallons). Ronald Edward Zupko, in his book *Revolution in Measurement* (1990), has compiled a hundred-page appendix for premetric weights and measures in Europe. Not only every country, but provinces, counties, and their cities had varieties of measurements by, so to speak, the bushel. Indeed, in some riven cities, one measure was used inside the walls — *intra muros* — and another was applied to transactions taking place in the unwalled suburbs — that is, *extra muros.* Moreover, different kinds of things were measured in different sorts of ways, even if the same word was used. In England, a "bale" of bolting cloth was 20 pieces, of buckram 60 pieces, and of fustian 40 pieces.

Interesting, you say, but what is this to me? And I suppose that is the question which has given rise to these reflections on what one might term "the metrification of mankind." I believe there is deep meaning in the progressive rationalization of our ways of weighing and measuring, but it is a meaning we singularly choose to ignore, in part because it is very difficult to confront the movement toward universal standards directly. What we are faced with is that rarest of occurrences: a gnarled and thorny problem which humanity has almost completely resolved. To cast our eyes over the myriad measures that once existed, to hear the pained measurement plaints that still resound out of the letters of grievances (*cahiers de doléances*) which the French people sent to their king at the beginning of their political and metric revolution, is to be removed from our present selves.

The collective cry of prerevolutionary France, "one king, one law, one weight, one measure," is simply illogical. For how can there be a world in which all meters are not 39.37 inches or, more precisely in the strict language of SI, equal to the distance light travels in $1/299,792,458$ of a second? The answer is, of course, that human affairs looked, and, even more important, felt, very different before the triumph of metrological science. And that is the first great truth to take from the twisted path toward standardization which we as a species have traveled over the last two hundred years.

Weighing and measuring used to signify much more than quantification. They were one of the most palpable and conspicuous manifestations of political power — and, by extension, their misuse symbolized the abuse of that power. To some, the concept of weighing and measuring itself came to represent the evils of all societies in which the poor seemed to be eternally shortchanged by the rich. Flavius Josephus, the acidulous Jewish historian of the first century A.D., tells his readers that the originator of metrology was none other than the world's arch-criminal. "The author of weights and measures, an innovation that changed a world of innocent and noble simplicity in which people had hitherto lived without such systems into one forever filled with dishonesty," was, he informs us, "Cain."

While the crooked merchant with his doctored scales and his shaved weights corrupted a part of the market, it was the lord in his castle who could pollute or purify all around him. It was he who kept the cups and weights which established the local standards. It was he who brought them out during the payment of harvest taxes and feudal dues and, by so doing, demonstrated his importance to the commonweal. Indeed, his influence could in some real sense be measured by the general acceptance of *his* measures. One clear sign of anyone's political impotence was his subjects' fealty to a different pound, mile, cubit, or barrel. Thus it is not surprising to find that, in medieval European towns where temporal and religious forces were contending, the churchmen had one measure, and the nobles another.

People also used the corrupted and the contradictory measures to mask something they understood existed but believed was quite simply evil. I am talking about those periodic fluctuations in the price of things that we call inflation and deflation. Merchants in medieval Danzig would cloak inflation simply by changing measures. Thus the size of a purported "pound's worth" of bread changed, depending on the cheapness of the wheat going into it, but the price stayed the same.

One should probably pause and reflect about this very other way of doing things for a moment. I would suggest that if some medievalist of a merchant tried to institute the elastic pound into his dealings today, his customers would think him not quaint, but deeply immoral. Price can yo-yo up and down, and inflation can explode into hyperinflation, but in our minds pounds and kilograms and hours are immutable. We imagine that to change them is in some way to change the laws of nature, and yet, only two hundred years ago, it was not clear at all that the only way to construct an economy was to keep the weights and measures fixed and fiddle with the money. Not clear at all.

To ask what finally standardized weighing and measuring is to be caught up in the contradictions of history. On the level of mass psychology, I am tempted to argue that we are where we are today simply because of one nation's collective bile — what the sociologists term *ressentiment.* The French, en masse, were unconditionally fed up with the confusion and inevitable arguments which flowed from their measurement pandemonium. Indeed, we can even quantify how displeased they were with their mismeasured lives. Among the various *cahiers,* those letters of complaint we talked of earlier, there were 727 calls for measure reform. However, the promiscuity of standards upset people for very different reasons: 110 complaints in the *cahiers* tied reform to a commercial benefit,

44 saw reform unsnarling legal arguments, and 98 linked the changes to a general hatred of feudal rights and privileges.[1] Moreover, while everyone knew something was wrong, what to do about it was still something of a muddle. National standards were called for in 440 *cahiers*, provincial standards in 69, and local standards in just 13.

Part of the confusion may have arisen because some of the French didn't want a measure revolution, but rather wanted a reinstatement of the old order — the very old order. There was a collective longing for that mythical moment in the past when the scales were all true, and every citizen meant exactly the same thing by a mile and pound. For a thousand years, various French kings had associated their calls for reforms in measurement with a return to a time when, as the historian Paucton wrote in 1780, "all measures were equal." That moment could be fixed. It was the ninth century when Charlemagne — he whose foot was to be *the foot* — set the standards no later monarch could enforce.

But there was also an important something else afoot, as it were. For while prerevolutionary France was in the throes of deciding how to banish its plethora of mismeasures, the European astronomers and chemists and physicists were demanding both more accurate and less human measures. Their rationale was quite simple: if you are going to have a study of nature rooted in the reproducibility of experiment, a common language of measure is essential. It seemed to follow that these measures should reflect the rational, universalist ethics of scientists and not the body parts and political intrigues of kings and commoners. So, for more than two hundred years, scientists had been searching for a nonpolitical measure. Picking up on a theoretically immutable principle first discovered by Galileo, the Royal Society of London proposed in 1660 that a universal length be equal to the length of a pendulum which would beat seconds. A different kind of inhuman standard was suggested by French mathematician Gabriel Mouton in 1670, when he proposed that the universal measure be the length of an arc of one second of the great circle of the Earth — that is, a circle like the equator which bisects the planet at its widest. This measure, the *milliare* or *mille* (roughly 30 m, or 34 yards), would be subdivided by tens into smaller units: the *centuria, decuria, virga, virgula, decima, centesima, or millesima.*[2]

In this, Mouton was following up on the suggestion first made in 1585 by Dutchman Simon Stevin that dividing things into units composed of ten things — *dismes,* he called them — brought sense into a premodern world in which most people's notion of number ended when they ran out of fingers and toes. In 1675, Tito Livio Burattini published a book in which he proposed that a "meter" — from the Greek word for measure — become the world's measure of length. He suggested that there was a desire on the part of "all civilized people on Earth to use the same weights and measures despite differences in language and customs" and that he saw it as his holy mission "to convince everyone to use weights and measures which were given to us by Nature, and which are as invariable (and untouchable) as long as the movements of heavenly bodies persist."[3]

The irony in all of this is, as we have seen before, that not only did no one agree with this rampant universalism, but most people of the time probably found it quite a bizarre notion. Before our times, metrology *was* in some sense who you were and where you lived.

Consequently, a reasonable compromise, if not rational metrological reform, was the progressive conservative approach which the British had begun to pursue. If you found the national measure standards had become a goulash, you simply standardized them. An inch became so long and no longer. A pound weighed this much and no more.

What the French and other scientists were proposing was much more radical. They were saying by implication that weight and measure standards were so important to their world that they could no longer be left in any ordinary person's or politician's hands. And, however one looks at it, they were also proposing a transfer of power.

We live in an age in which the triumph of scientific thinking is so complete that we are barely aware of the historical universes it has snuffed out. All cultures once used their version of Annas as units of measurement until each one was banished by a scientific ideology which — to turn the Greek saying on its head — makes all kinds of invariable things the measure of man.

And that is followed hard upon by a very contradictory truth. For almost two hundred years, the rush toward metrological standardization has represented the French way of reshaping the world — the all-or-nothing way; the revolutionary way; the scientific approach to understanding nature. But, while the metric system was set in motion by calls for reform, what really established it was an idealism so single-minded that even after two centuries it is difficult to believe its adherents were not just crazy.

After the French Revolution began, and after all the disgruntled petitioners had sent in their *cahiers*, those newly installed in power decided they would respond. This itself was notable, as a variety of prerevolutionary reformers had heard the cries for metrological change but, after considering the scientific case for a total revision, had pulled back. Weight reform and social reform went hand in glove, and it was a glove which pinched many of the king's faithful servitors. However, if feudalism and yearly payments and unjust contracts were going to be abolished, then nothing stood in the way of drastic alterations in weight and measures. So, in April 1790, the Bishop of Autun, or, as we more familiarly know him, Talleyrand, brought the issue of a root-and-branch reform before the National Assembly. He proposed that a joint committee be established by the French Academy of Sciences and the Royal Society of London to ensure the new system had a truly scientific scope.

Once the proposal for measurement reform became law, a committee of the French Academy took up their cudgels and looked at a variety of ways in which it could be accomplished. The English turned down the arrangement as "impractical" and, in so doing, "probably reasoned," says measure historian Ronald Zupko, "that cooperation on this metrological venture would give tacit approval to republican political upheavals and bloodshed."[4] Despite the coolness of the English, and after debating the merits of a system based on units of 12, the French Academy ultimately recommended that the new system be decimal.

Their plans included a proposal to divide the year into 12 months composed of 3 weeks, each of which were to be 10 days long. Each day was to be divided into 10 hours, each hour into 100 minutes, and each minute into 100 seconds. The "deciday" (2.4 hours), the "milliday" (86.4 seconds), and the "microday" (0.864 seconds) were met with intense opposition in a clockwork world in which the medieval people's penchants for hours of different lengths had already succumbed to standardization. Since a standard was already in place — albeit one which harkened back five thousand years to ancient Sumer and a Mesopotamian numerology which grouped those things to be counted into easily divisible units of 12 and 60 — the French metrological revolutionaries were forced to retreat. Everyone wanted a standard weight and measure, but nobody wanted to throw away the old clocks and watches and time-worn habits.

At the same time that they were considering the decimal day, the French Academy decided to make a meter equal to one ten-millionth part of a quadrant — that is, a fourth — of the Earth's circumference. The question then immediately became: How much is that?

There were previous estimates, and indeed trying to measure accurately the circumference of the Earth had become something of an obsession in the age of science and exploration.

In 1525, Jean Fernel had ridden in a carriage from Paris to Amiens with a bell mounting that sounded each time a wheel completed an entire revolution. From his rolling data he had produced a crude measure of the Earth's girth, but it and subsequent remeasures weren't good enough for the radical instincts of 1792. The Revolution's new measures not only had to be derived from nature but also had to have an immutable quality, a state of being the *philosophe* Condorcet described as "For all people, and for all time."

So an exhaustive scientific resurvey had to be undertaken. The basic principle underlying the calibrations were those laws of angle and side which students today learn in high-school geometry and trigonometry. The length of all sides of a triangle can be determined if the length of one side and the size of two angles are known. If you measure a variety of triangles sharing a common side, you can determine that side's length with great accuracy.

In practice, triangulating France entailed a complicated series of both horizontal and vertical measurements, which had to take into consideration: (1) changes in altitude, and (2) the linear distortion caused by the bending of straight measures over the Earth's curved surface. To produce the triangles, reference points were needed, and so scaffolds had to be built, and the heights of church towers measured with as great an accuracy as possible. As well, constant star readings would be necessary to ensure that surveyors knew where the meridian — the imaginary line of the Great Circle they were tracking — lay.

In order to arrive at this measure accurately two astronomers, Jean-Baptiste Delambre and Pierre Méchain, were asked to survey a little more than a tenth of a quadrant — 9⅔ degrees. This was the distance from Dunkirk on the English Channel to Barcelona in Spain.

In the last week of June 1792, plum in the middle of the French Revolution, Méchain and Delambre left Paris. Méchain headed south, and Delambre north, to make the definitive measure and produce from it the definitive meter. How difficult the task would prove was demonstrated that first day. Méchain had not even reached the suburbs of Paris before he was arrested by a Revolutionary guard trying to catch aristocrats fleeing the city. His crime was that his letter of passage did not specify what the various surveying instruments he carried were to be used for, and none of the guardsmen or passersby could understand the astronomer's explanation of their application. Who can blame the doubting Parisians? If ever an explanation sounded as if it had been cooked up by the devious to hide more sinister purposes from the gullible it had to be Méchain's claims of an Earth survey.

The arrest and subsequent freeing of Méchain and his assistant signaled the nonscientific problems the astronomers would face during their calibrations. Once the surveyors were stopped because the sheets they were using as survey targets were white, and white was the color of Royalists. Clearly some counterrevolutionary plot was in action. In the middle of his calculations in 1793, Delambre was forced to return to Paris and turn in his surveying equipment when the Terror started executing the country's scientists. Both Lavoisier and Condorcet, who had been members of the provisionary weights-and-measures committee, were killed in the name of the people, and Delambre himself lived for the better part of a year in the shadow of Madame La Guillotine. And while surveying in Spain, Méchain was injured so severely during a visit to a water mill that, for a while, it was reported he was dead. Once recovered, he found himself barred from returning home for several years because of the fighting which had broken out between Spain and France. In addition to the chaotic political situation, foul weather, scrambles up and down mountains, and illness abounded.

Ultimately, after six years of surveying and the laying out of more than a hundred overlapping triangles across the face of France and Spain, the *mètre définitif* was arrived at in 1798. The next year it would replace a provisionary meter in use since 1795. In so doing, it, and the gram which represented the weight of a cubic centimeter of water, officially laid to rest, in France at least, the economies of mismeasure I described earlier.

Well, perhaps I go too far. For there is an ultimate irony in my brief retelling of the tale of mad measuring in the midst of political chaos. The provisionary meter was based on an attempt by Cassini fifty years earlier to measure the Earth's circumference. After all of their efforts, the intrepid French astronomers had apparently shown that the provisionary meter was a third the thickness of a nickel — 0.325 millimeters — too long. However, a later recalculation showed that the *mètre définitif* was in fact too short by 0.23 millimeters. And so, after all Méchain and Delambre's travails, it turned out that Cassini's provisional meter had been a more accurate measuring rod than the definitive one which had been cast to their heroic standards.[5]

I don't mean to imply in this that the metrological pilgrimages of the two triangle-crazed astronomers were ultimately only a joke. To the contrary, the tiny differences between the right and righter meters are vastly more important than they seem. They are palpable expressions of the divisions between science and society which define modern times. At the start of the French Revolution, it appeared that society had hired science to put an end to the chaos of mismeasure. From our perspective, it is clear that the measure revolution had been hijacked by what amounted to a new faith. The interests of the merchants, the consumers, and the scientists were not congruent because the scientists and engineers wanted a precision which transcended the needs of the ordinary person, who wished only that his or her cheese and wine and bolts of cloth would be dispensed according to a uniform standard. Speaking as a politician, Condorcet himself would complain about the meridian measurement's overzealous effort "to provide a perfection which is not absolutely necessary."[6] What he might have more aptly said was that it wasn't absolutely necessary if you were not an adherent of the new religion of analysis in which weighing and measuring had taken on a transcendent form. I think that, from a nonscientist's perspective, the only sensible way to look at Delambre and Méchain is to view them religiously. However mad the enterprise now seems, they were in fact pilgrims and holy men. It was their sense of the holy that allowed them to ignore the chaos in which they found themselves. However crazy their efforts, it was a holy crazy, undertaken in the name of a scientific methodology which transcended its doers — and maybe even its species. And just as surely, the travails of Delambre, Méchain, and their fellow metrologists revealed a blood truth about the scientific-industrial theology which was to follow: their pilgrimage to map the meridian demonstrated that the holy grail of scientific man was not something natural he worshipped, but something unnatural he made — a meter stick, a light bulb, a telephone, a rocket ship. And it was a holy grail built on a fanatical weights-and-measures rigor. Even without a computer or a laser to aid them, the new measuring class's ability to hone their devices was astounding.

It is also astounding how much metric's growth resembles the various conversion histories which previous universalizing Western religions passed through when converting heathens. Sometimes — think Christmas trees and Easter eggs — there have been attempts to graft the new religion onto the faith of one's fathers. Even in metrology's homeland, allegiance to metric initially wavered, as French men and French women tried to reconcile the familiarity of the old and homey with the shock of the authoritarian new measures. In 1812, with the didacticism of the Revolution on the wane, the French took a step backward and returned to what were called "usual measurements." A metricized *toise* and the *livre* and the *aune* and the *boisseau* were allowed. This proved confusing, however, and in 1840 the country reverted once more to the true church of kilogram, meter, and centigrade.

Like other great Western religions, metric has its own holy city, or rather, holy building. The white-shuttered Pavillon de Breteuil, once a royal palace situated at Sèvres, outside Paris, has been designated an "international enclave" and is not subject to French law. It is there that the sacred, metrological relics, the physical meter sticks and platinum iridium kilograms, are stored and protected from all ravages of time. (See "The last metric artifact.")

But the most striking feature of metric's conversionary process has been the resistance of the old religions. Even though England had been asked to join at the very beginning, and the native American *philosophe* Thomas Jefferson proposed a decimalization of the English weight system in 1790, an ecclesiastical war followed the birth of metric. Initially, much of this undoubtedly reflected a continuing revulsion with the excesses of the French Revolution and its deification of reason over religion. The ten-day-week metric calendar had not made provisions for a Sabbath, and Christians in the United States in the nineteenth century found this proof that the system was essentially an abomination. One pamphleteer could look at a meter stick and see a cloven hoof: "This system came out of the 'Bottomless Pit'," he wrote. "At that time and in the place when this system sprang it was hell on earth. The people defied the God which made them; they worshipped the Goddess of reason. . . . Now, my friends, when the grave diggers begin to measure our last resting places by the metric system, then understand the curse of the Almighty may crush it just as he did the impious attempt to abolish the Sabbath."[7]

If metric was atheistical and sacrilegious, then, conversely, English measures had to be holy. They, said proponents of the inch, foot, and pound, grew from the cubits a francophobe God gave to Noah to build his nonmetric ark. And in the 1880s there arose in the United States the International Institute for Preserving and Perfecting Weights and Measures, which stated that it believed "our work to be of God's . . . [and] we proclaim a ceaseless antagonism to the Great Evil, The French Metric System."[8] It further found the inch to be the base measure of the Great Pyramid of Cheops, a discovery which proved again the English measure was divinely blessed.[9] One can only imagine the disorder which broke out in Heaven when, in 1978, and some years after England had ceased being a pounds-and-foot country, a new translation of the Bible converted measures to metric. Goliath, formerly "six cubits and a span tall" and arrayed in "500 shekels of brass," turned out in a SI world to be "nearly three meters tall with 57 kilograms of armor."[10]

By 1904, businessman Samuel Dale was suggesting that those who advocated metric were in fact arch-enemies of the American civic religion: The Unfettered Pursuit of Whatever Anyone Wanted to Pursue. "We have no king to order a change of our standards . . . of weights and measures, no established church or aristocracy to execute royal decree. In the place of a people accustomed to being controlled by an arbitrary government, we have a people who govern themselves, and who are quick to resent the interference of the police power in their private affairs."[11] As you might suspect, this argument was later expanded to include communism — Russia became metric shortly after the 1917 revolution — another of the un-American authors of metric compliance.

But the spirit of the Anglo Saxons' resistance to metric otherness can be coalesced in one unforgettable utterance in the 1920s, when Prime Minister Lloyd George interrupted a member of Parliament speaking in favor of a change to metric. "Would you wish that an English workingman going to a pub to drink a pint of ale be required to ask instead for .56825 liters of beer?"[12] he thundered. It appears God had specifically reserved inches for the Egyptian pyramid builders and pints for English tavernkeepers.

By the late twentieth century, the English measure forces would leave off portraying their standards as God-given and instead suggest that a devilish economic plot had been hatched. European businessmen were using metric, and the costs of changing to it, as a way of subverting their competitors. "If we allow them to coerce us into shouldering a $100 billion or so metric conversion cost it will seriously aggravate our already excessive inflation, raise our production cost and therefore allow them to undersell our manufacturers by even greater margins than at present," wrote Vermont engineer J.W. Batchelder in 1980.[13]

If there has been a holy war fought in the name of weights-and-measures standardization, then we must also say that a kind of metrological monotheism has triumphed. Ultimately, again making an exception for the United States, the conversion process has brought all of us under the canopy of metric's decimalized tent. We have been forced to find different ways of expressing the local, regional, and national traits which two hundred years ago exhibited themselves in the multiplicities of pounds and feet and bushels.

Metric has triumphed for what are probably self-evident reasons. We are all part of a science culture in which everyone trades and makes goods for everyone else, and in which basic principles of physics and chemistry and biology are the same for all people and all time. We all believe, as a matter of faith, that changes in the value of currency and not in the value of weights and measures are the true variables of economy.

We are one weighing-and-measuring church, but that church's pews do not provide an altogether comfortable fit. Metric is uniform, and its rigidity has foisted a uniform imperfection on the world. A gram is really too small to weigh anything but a single paper clip or a large raisin. Counting by tens is of course a rational way of doing measures but, as metric opponents have pointed out endlessly over the past two centuries, it often flouts its precision in the face of the lower level approximations of everyday life. Asking for two and a half pounds of sugar seems a more natural statement than requesting 1,500 grams of anything. Doubling or quartering of measures appeals to something quite primal in the human brain.

It ties into a mental arithmetic where the thing to be reckoned is lumped into gross categories of "bigger" and "smaller." Metric eschews vulgar fractions and, in so doing, hides another truth in the triumph of the universal measure: we now count things in ways that don't reflect how we think about them. And that can be quite clearly demonstrated by what everyone recognizes as the most common mistake one makes in applying a metric measure. The easiest stumble is to be off by an order of magnitude. Can anything be more odd? In the name of a universal precision we have allowed ourselves to be ten times more dramatically wrong than in the past. The vision of an absurdly inaccurate accuracy seems a fitting end to this little reflection on the two-hundred-year revolution in measurement and the ultimate installation of the metric god.

But in the last few days I have found myself looking again at the door frame on which Anna's growth line had been marked. And I realized that there was another measure there which I hadn't noticed initially. The door itself measures us. It is constructed to our heights, and it reaches its limit regardless of whether a meter or a yard or *toise* is laid against it. Similarly, however we define it, a kilogram is still something you can carry easily in one hand, and a few degrees Celsius is still a unit of temperature change you can feel on your skin. But there is nothing which suggests that our universal measures are indeed the measures of the universe. If and when we encounter alien others, one of the most alien things about them may be their very, very inhuman weights and measures. They assuredly won't use our hours, days, and seconds with their rootedness in the length of an Earth year. Their basic measures may be thirty times heavier or a hundred times longer than our body metric is comfortable with. And what would it be like if, in the name of true universals, they asked us to change over to a measure which had no human consciousness embedded in it? Could we? Would we? Or are we limited in our ability to be universal by the same assumptions which underlay our ancient and promiscuous former measures? Is the meaning of all this standardization ultimately that, no matter how fine we cut the cloth, hands and fingers and yells and jumps are the only yardsticks humans truly feel comfortable with?

That thought comforts me in ways I do not understand, so I have decided that, when I finish writing this, I will go upstairs and pencil a note on the wall to whosoever might come to live in this human-made, human-measured house after me. It will say: An Anna, the ultimate human scale.

The last metric artifact

The kilogram stands alone.
The kilogram stands alone.
Hi-ho, the metric system,
the kilogram stands alone.
(With apologies to nursery rhymes everywhere.)

The kilogram does stand alone. All the other units of the metric system have been redefined into nonhuman components. A "meter" is the distance light travels in $\frac{1}{299,792,458}$ of a second, and a "second" is the duration of 9,192,631,770 cycles of radiation associated with a change in the energy level of a cesium atom. But Mr. Kilogram remains, in the language of measurement, an "artifact." All the kilograms in the world are copies of the primal kilogram, which is made out of platinum and iridium and kept in a vault at metric central, Le Bureau International des Poids et Mesures (BIPM), which itself is housed in an old palace in Sèvres, near Paris. Size-wise, the kilogram is about the shape of the largest piece of salami a big man could slip into his mouth. It is housed under an old-fashioned bell jar, which itself is housed under another bell jar. Together they look as if they have been lifted from some grade-B movie's idea of what the Philosopher's Stone should look like. That stone was, in case you don't remember, a material that had the ability to change lead into gold — a "transmuter," in *Star Trek*–speak.

The irony, of course, is that the kilogram has not been able to be transmuted. It remains our measure because, frankly, nobody has been able to come up with a nonhuman equivalent. It's odd in a way, because originally it wasn't human at all. A gram was supposed to be the weight of a cubic centimeter of water at its densest point, which itself was supposed to be the temperature at which water froze, 0°C. Unfortunately, it turned out that water actually was densest at about 4°C, and even more unfortunately, that different purities of water produced varying-sized grams. Therefore, in

1889, the measurement community adopted the international prototype kilogram as its standard.

What that means is that countries wishing to adhere to the standard get a copy of the Paris-based kilogram. The BIPM guarantees these copies will be accurate to within plus or minus forty parts per billion. Nations use their platinum-iridium copies to calibrate various reference standards, usually made of stainless steel or brass, which are then used to check scientific and industrial scales. Once every ten years, the national prototypes are compared with BIPM prototype copies. As you might suspect, the result of all this copying from copies of copies is a slight error rate. Even though they are handled with tongs covered in lens paper, the national standards increase in weight by about 1 microgram — one part per million — every year. And when the comparisons with the steel and brass kilograms are taken, the error rate is about 30 micrograms, not a lot for you and me (about 1/10,000 the weight of a medium-sized ant), but an enormous problem for scientists used to measuring things with accuracies of parts per trillion. "It is exceedingly unlikely that the international prototype kilogram has the same mass as when it was first adopted as the unit in 1889, or even from one comparison to the next," wrote a distraught U.S. physicist in the journal *Physics Today* in 1993.[1] That effectively reduces the theoretical accuracy of any international mass comparison to, perhaps, ten parts per million.

Beyond the eensy-weensy difference question, there is a larger political problem. What happens if something calamitous occurs to the Paris standards, where several "originals" are under bell jars? As George Chapman, of Canada's National Research Council's Institute for National Measurement Standards, clearly put it, "If someone drops a hydrogen bomb on Paris, it is going to be a real pain in the ass for those of us who have to pick up the pieces."

And so we may yet see in our lifetime a revolution in kilogramage — the end of the last artifact in measurement. Proposals are afoot to redefine the kilogram. Some suggest that it be equal to a specific amount of pure silicon. One would be able to predict from the physical nature of silicon that a cube roughly 7.5 cm (3 in) high should weigh exactly 1 kg. Others have suggested that the nonartifact kilogram be defined as the amount of electric current required to create a magnetic field that would keep a kilogram in balance on a scale. In this measure, the kilogram would be defined as something like 30 volts of electricity passing through a wire of a certain size.

The end result would be a metric measuring system entirely based on abstract physics. While this might not be a Big Bang in your personal life, it is remarkable to consider that, with the ejection of the artifact kilogram, humans would have constructed a scale that has nothing human within it. Not only does this make me unaccountably wistful for the days when people were the measure of all things, but it also makes me wonder: Is my sense of dispossession anything like what God felt when he first saw humans making things which never existed in nature?

The earth-quake-saurus

Every time there is an earthquake in California, I stare at my fingernails. It is about the only time I have anything like a deep communion with those hardened hunks of clear, dead tissue. As the television shows pictures of bridges collapsed and cars crushed into slabs, I regard those keratin-filled first cousins of hair and silently intone to myself a science writer's mantra:

SanFranciscoismovingnorthwardatthespeedfingernailsgrowinayear.
SanFranciscoismovingnorthwardatthespeedfingernailsgrowinayear.

To tell you the truth, I don't know who must be credited with first coming up with this highly evocative way of describing the slow but inexorable movement upward of a part of the Pacific plate on which California sits. However, every geophysicist in North America now seems to be aware of the measure and gives it in response to that inescapable question from reporters: "How fast does that plate move and can you compare that to anything, like, human?"

In and of itself, the fingernail imagery is quite wonderful. It has a certain Zen, beach-bum quality. Say:

SanFranciscoismovingnorthwardatthespeedofafingernailgrowinginayear

eight times quickly and you can almost hear the existential surf crashing inside your head. A drifting continent has all the unstoppability of a fingernail.

A wonderful image, but one which quite early in my reverie I also realized that nobody else watching California rock and roil is truly interested in. We are not a metaphorical and Zen culture. We are a scientific culture, and we expect to have earth movement conveyed precisely; we want to have the fingernail measure turned into a number, something which fastidiously describes the mayhem that occurs each time a plate moves. And when I look up from the Zen of my fingernails, I realize that the television reporter will soon tell me just such a figure. I mean, of course, the seemingly simple scale of Mr. Charles Francis Richter and his portentous 7.6s and awesome 8.2s and god-forbid 10s. The reporter will use the number and assume that I will understand the Richter scale as easily as I did the fingernail image.

But I won't. Not intrinsically. Not historically. Not truly comparatively. I won't, because I never know what the Richter numbers mean in terms that touch me.

Nonetheless, I will listen, nod my head, purse my lips, and disguise my confusion, because Charles Richter's rating of earthquakes is one measure every modern person is expected to comprehend instinctively. Indeed, in sixty years, Richter's numbers have so completely entered popular consciousness that they have themselves been turned into a scalar metaphor. The economic fallout of a megamerger is not just big, but "off the Richter scale." Excitement about Toronto winning its second World Series is rated a "10 on the sports Richter scale." And because for the longest time I assumed that everyone else was completely comfortable with Richter's measure, I thought that I was doomed to fake understanding forever. It was the sort of thing cynical friends could have engraved on my tombstone: "Here lies Stephen Strauss. He was the sort of person who knew 7 was a lot larger than 6 on the Richter scale, but was afraid to admit he didn't know why, or how much bigger." Then a kind of salvation occurred while I was browsing through the pages of *Nature* magazine. There on page 512, at the bottom of an article about the likelihood of Seattle being riven in two by a huge earthquake, I came upon the startling words of U.S. geologist Thomas Heaton. "The many different existing magnitude scales are generally all included together in the maddeningly vague term 'Richter scale.' This is popular with the press, but meaningless to a seismologist," he wrote.[1]

Hooray, I thought, I have a right to be confused about degrees Richter — even the scientists are. And with that, I vowed to one day try to make sense of Richter's scale. Simply put, I wanted to know: Why is something which looks as simple as 1-2-3-4-5-6-7-8-9 so bloody confusing?

What I have learned is that part of the answer, as with many measures, emerges from the history of the scale and the awkward interaction of science and technology which underlies it. Or, to put it in a homelier fashion, earthquake measurement follows the old adage of "He who has a yardstick measures the universe in yards." The Chinese were the first to calibrate earthquakes, using the person-sized, wine-jar-shaped "seismoscope" built in 132 by Chinese astronomer royal Zhang Heng. However, it was a very limited measuring device. The direction of a quake was indicated when a ball dropped out of the mouth of a bronze-cast dragon into the mouth of one of eight metallic frogs beneath it. And, as with many Chinese technological breakthroughs, what developed out of the seismoscope was not science, but mechanical awe. The Chinese didn't ask if there was a pattern to where the earthquakes occurred that could be quantified, or whether it was possible to measure the strength of the quake. No, it was the seemingly magical relationship of the earth movements and the ball drops which transfixed Chinese observers. When the seismoscope was able to record an unfelt earthquake which occurred 650 km (400 mi) away, the court historian recorded "everyone admitted the mysterious power of the instrument."[2]

At the beginning of the age of science, the Europeans were also creatures of awe, but their reverence was increasingly reserved for quantification. They were (and we still are) entranced by Descartes's observation that nature is "a universal mathematics." A true understanding of nature came not through wisdom, but through systematic counting. And, in that line, Italian physician Domenico Pignataro began some of the demon numeration which separates a scientific culture from all cultures which existed before it. In 1786, he classified 1,186 quakes which occurred in the

Calabria area of Italy between the years 1783 and 1786. These he divided into categories of slight, moderate, strong, very strong, and violent. In the same report, the chief physician in the court of Naples, Giovanni Vivenzio, translated these categories into a more humanly meaningful number. He outlined the damage and death which occurred in villages affected by these quakes, and therein illustrated the dualism intrinsic to quake quantification. The ground moves, and that movement smashes things up. It was not clear then, and in many ways is not clear now, which is the cataclysm's most important feature. Today's seismologists are interested in what the tremors tell us about the nature of the Earth; today's engineers want to know what kind of buildings to build to avoid the earthquake damage; ordinary folk want to understand the extent of their danger and a quantification of that havoc. But, in the beginning, there was no simple way of separating the shifting ground from the damage it caused. Thus, earthquake measures unselfconsciously annotated the quake's effects on people and their belongings. Accordingly, in 1828, P.N.C. Egen, from his study of a Dutch quake during that year, suggested a six-point intensity scale. Number 1 was a quake that you could miss if you didn't concentrate. Number 2 saw a few people feeling the shock, a few small potted plants vibrating, a few glasses that were close together jingling. With number 3, the windows rattled, bells in houses rang, and most people realized there was actually an earthquake going on. Number 4 saw the furniture shift a bit and left no doubt in anyone's mind that the earth had moved. In number 5, the furniture was strongly shaken and walls were cracking, but "only a few chimneys [were] thrown down." And finally, in category 6, furniture was strongly shaken, mirrors and glass broke, chimneys tumbled down, and walls cracked.

Simply put, Egen's scale measured how quakes interacted with nineteenth-century northern European housing technology and communal organizations. Egen's measure also looks decidedly

Dutch. The worst consequences were moderate and middle ground, as suits a country where villages weren't regularly being razed by the earth's movement. By the end of the century, several varieties of intensity measures came into being which stretched the scale to include more horrific consequences. The Italian Michele Stefano De Rossi and the Swiss François Alphonse Forel combined to produce a ten-point scale which related quakes to such things as clocks stopping, chandeliers shaking, general panic, and "a general ringing of bells." Number 10 consisted of "great disasters, ruins, disturbances of strata, fissures in the earth's crusts and rock-falls from mountains."

By the early part of the twentieth century, intensity became, as best there was, the world measure of an earthquake. Modified into a twelve-point scale by priest-geologist Giuseppe Mercalli, it has served as the basis of a variety of locally tailored scales. Two Americans in 1931 included Southern California–type references to "steering of cars affected" and "concrete irrigation ditches damaged." New Zealand threw in mentions of damaged domestic water tanks, which were found on many of its farms. To accommodate individual differences in 1958, two Russians calculated there were forty-four different intensity scales at work. However, despite all efforts to customize it, the Mercalli was a scale that everyone understood was seriously flawed. How accurately was anyone measuring a quake's intensity? Confusion often existed between what the media said had happened and how an intensity rating was "scientifically" calibrated. "Newspaper reports are useful," wrote Charles Richter himself, "if one becomes accustomed to reporters inserting details which 'ought to' be there whether they correspond to fact or not. 'Buildings here were shaken' means only that persons felt an earthquake and does not imply any structure vibrated visibly."[3] And what about quakes which happened under the sea or in unpopulated areas? Were they somehow unauthentic because no human was there to measure and observe the damage in accordance with the

Mercalli scale? How were alternative building technologies to be compared? The Chinese, who had been recording quakes for three thousand years, regularly made mention of catastrophic fires in describing the effect of their huge tremors. And what about the fact that locally sharp quakes were given a higher profile than quakes that affected a larger area?

In this turmoil, Charles Richter, a compulsive seismologist who taught at the California Institute of Technology, began to put a scientific face on earthquakes. How compulsive was he? He was a man so obsessed with earth movements that he had a seismograph installed in his house and regularly commandeered the one telephone in the Caltech laboratories so he would be the voice of authority to the press. But in the early 1930s, he was beginning a career and facing a very much smaller problem than how to construct a true scale to compare all the world's earthquakes. What he was trying to do was prorate the two hundred to three hundred quakes that took place in Southern California every year. The reason for this was that Caltech had begun issuing monthly earthquake reports from the seven seismometers it had distributed around Southern California. The feeling at Caltech, particularly after the panic which had been associated with the 1925 Long Beach quake, was that these reports had to be quantified in some easily understood manner for the general public. "We felt a certain responsibility to keep the public informed, particularly as misinformation was often seized upon and twisted in a way which was contrary to the public interest," Richter later recalled.[4] It is important to dwell for a second on this original motivation. The scale was a blatant public-relations effort. It was for us, and not for seismological science, that Richter was initially working.[5]

All Richter had originally hoped for in a public-measured science was a very rough approximation. An earthquake scale of "small, medium, or large" was his original goal. To achieve it, ways of comparing seismographic readings that were being taken from seven locations at various distances from the quake's center had to be devised. To appreciate the difficulty, you have to know a little about earthquake waves and seismographs. An earthquake is a harmonic; that is, what we feel and see is in some sense the various ways in which the energy from a sudden shift in the Earth's crust vibrates the instrument we call planet Earth. When the ground suddenly shifts, it generates a series of very different waves. The first to appear are P (for primary) waves. They alternately compress or expand rock, lava, or any other substance in their path. The roaring sound which often signals the onset of a big quake is the result of P waves that have reached the surface and are agitating the air.

P waves are followed by slower S (for shearing) waves, which jiggle rock particles from side to side and up and down in a contorted movement we might liken to a dirty carpet being simultaneously shaken and stretched. Collectively, P and S waves are called "body waves," because they move through the body of the Earth. But there also exist surface waves of ground motion, Rayleigh (R) waves, which move across the surface like rolling ocean waves, and Love waves, which have a side-to-side slithering quality. As a general principle, a seismograph uses weighted pendulums to make measures of all these different waves. The pendulums are suspended from a frame that is itself anchored to the ground. During a quake, the movement of the weighted pendulum lags behind that of the frame. The pendulum's relative motion is then recorded — at its simplest, by a needle scratching a line on sooty paper.

The problem for Richter was trying to figure out how the various wiggle patterns could be fitted together. He made a breakthrough when he came upon a paper by Japanese seismologist Kiyoo Wadati. Wadati described a rule by which a seismic wave diminished in height the farther it traveled from the quake center. This would allow readings from a variety of seismograms to be compared if the epicenter of the quake was known. Richter was able to verify Wadati's demonstration of wave diminishment, but the "corrected" values didn't lend themselves to an easy scaling. Collectively, when the smaller quakes were put at the bottom and the bigger ones at the top, a kind of chevron pattern (something like a sergeant's stripes) was produced. To give you a sense of the scaling variation, the smallest amplitude reading was about 1 millimeter — half the thickness of a dime — and the largest was about 12 cm — about the width of my hand.

Richter was stumped about how to make something meaningful out of these differences; then Beno Gutenberg, who directed the Caltech seismic laboratory, suggested he arrange the quake readings logarithmically; that is, set them in a scale in which the differences between quakes would increase by ten times for each number on the scale. Logarithmic organization had already been done for a variety of scientific measures, ranging from the brightness of stars to the pH levels that depict how acidic or basic a substance is. Not only did logarithmic progression allow the quakes to be compared, but it gave the immensity of the differences between them a clear face. As Richter would later comment, "If there was anything you could call an actual discovery that came out of the scale, it was that the biggest earthquakes were enormously bigger than the little ones."[6]

Richter announced his computations of what he called "the magnitude scale" in a 1935 paper. The magnitude scale was initially very localized, being tied to readings from Caltech's special type of seismograph. It worked only with relatively shallow — no more than 16 km (10 mi) deep — earthquakes, and it was applicable only to quakes occurring within 600 km (400 mi) of a seismographic station. The standard magnitude for a given quake was one which a Woods-Anderson seismograph might have recorded if it was located about 100 km (60 mi) from the quake epicenter. This became known as the "local magnitude" of a quake, or M_L in the shorthand of seismology.

In his original paper, Richter went so far as to suggest how his new scale might have applied to some previous large quakes, and estimated that the 1906 San Francisco quake was probably between 7 and 7.5 on the magnitude scale. "How far above this the magnitudes of actual earthquakes may extend is a difficult, and in one sense an unanswerable, question. Judging by the relative amplitudes of distant records shocks, there must be cases of at least magnitude 9, and very probably 10," he wrote.[7]

This reference was the origin of the notion of the infamous "10 on the Richter scale." In truth, as Richter emphasized, the scale is open-ended and arbitrary. It goes as high — or as low — as nature permits and as machines can measure. As well, Richter took great pains to point out in the original paper that the zero which started the scale was not an indication of "no quake," but rather a baseline. It was only the smallest earth tremor that the seismographs of the day could detect. And indeed, since Richter's era, instruments with more refined sensibilities have been developed, and seismologists now report recordings which baffle the public by registering as –2 on the Richter scale.

Gutenberg immediately realized that the magnitude scale had possibilities as a worldwide standard if it could be detached from its Southern California biases. With Richter's help, he proceeded to universalize the measure, and almost as quickly made it more complicated. Richter's initial scale had been rigorously minimalist. It computed a number based on the maximum line height recorded on a seismograph of a given type. However, everyone knew that this was a wildly artificial thing to do. In the original Richter calculation, absolutely no distinction was made as to which wave was being measured, and even Richter admitted that, depending on where you were in relationship to the epicenter, almost any wave could be the highest. While wave discrimination didn't matter if all that was being measured, using the same machines, was shallow Southern California quakes; deeper and farther was another question. This was soon addressed by turning the Richter scale into the Richter scales.

Theoretically, one could make a scale out of the seismic fingerprint produced by each wave type. In 1936, Gutenberg and Richter published a paper in which a standard measure using horizontal surface waves which occurred over a 20-second interval was used. This spread the distance at which the scale was applicable to 1,000 km (600 mi) and allowed varying types of seismographs to be used. It too got a scientific nickname — M_S for magnitude of teleseismic (distant) earthquake.

Ideally, the two measures should have collapsed into each other; that is, the same magnitude should have appeared no matter which wave was chosen to be the yardstick. Alas, earthquake measurement is not a domain in which the natural phenomenon respects the measuring implement. When measuring quakes below 5 on Richter's original scale the M_S was often as much as half a magnitude off. Half a magnitude is about three times smaller, a magnitude paycheck of 33 cents on the dollar.

Besides, Richter scale two did not permit a computation of the magnitude of deep quakes either, because these shocks did not generate surface waves and because, when they did reach the surface, the surface waves' amplitude was seriously dampened by complicated changes in the Earth's crust. Therefore, Gutenberg came to believe that the most accurate way of assigning magnitude to deep and extremely powerful quakes was to calculate the magnitude of body waves. What emerged out of this was yet another Richter-type measure that lumped together P and S waves and was called M_B. By the 1960s, when an earthquake occurred all of the dreaded M measures — M_L, M_S, M_B, and scales that combined M_S and M_B averaged in some way — were being gathered and reported, often without explaining which was being used. And major disputes were breaking out between the founders. Gutenberg came to believe that M_B was the most accurate way of assigning magnitudes and, in his publications, reported all magnitudes in that dimension. Richter thought Gutenberg was wrong and was sowing confusion. He reported his findings as either M_L or M_S, arguing "in many instances it has been shown that initial waves are those of a small foreshock, to which alone the magnitude supposed determined for the following shock will apply."[8] By way of example, the huge 1964 Alaskan quake registered 8.6 on the surface-wave Richter scale and only 6.5 on the body-wave scale.

As scientists began to realize the importance of the scale to their, and not the public's, understanding, they cheerfully spooned in complications to Richter's oh-so-simple initial "big wave means big quake" rating system. The Mercalli intensities were roughly correlated with various Richter measures. This regularly caused journalists and the public to confuse the one with the other and report that a quake was an 11 on the Richter scale when what they meant was an 8 on the Richter scale (which results in a 12 on the Mercalli scale).

But this is child's play when compared with Richter and Gutenberg's computation of the actual energy released during a quake, which appeared the year after the first paper. After a variety of computations, the two seismologists determined that, for each number on the Richter scale, the actual energy released in a quake went up not by a factor of 10 but by one of 31.6.

This means that the "Richter scale" — whatever that might mean to an ordinary person — has two separate and distinct logarithms embedded within it. The difference between a 6 quake and a 7 quake is *both* 10 in magnitude *and* 31.6 in energy release. Since energy release seems so much better a description of what happens during an earthquake, you might well wonder why it has not simply displaced the magnitude measure. Richter had an explanation.

"Frequently there have been suggestions that the scale should be defined in terms of energy, but to do that would have involved continuous revisions, both numerical and theoretical. I have always insisted that the magnitude scale represents what we observe, and this may or may not be interpretable in terms of energy," he said.[9]

And indeed his doubts have been borne out, as over the last fifteen years or so there has been a massive revision of the energy releases of quakes. Scientists realized that at the top end of the magnitude scale — above 7 or so — there was a blurring of differences between quakes. A quake which was the result of earth movement over hundreds of miles could get the same reading as one which was very localized. As a result of the revision, a 1960 Chilean quake was upgraded from what had been an 8.3 on the traditional Richter scale to a 9.5. Later calculations would suggest that this meant it had produced between 35 and 40 percent of all earthquake energy released between 1900 and 1989.[10]

While this is all fascinating for the professional, where are we in terms of Richter's original noble purpose — you remember: a public-relations effort to make earthquakes comprehensible to us common folk? Well, I'd have to say that Richter's chevroned seismograms ultimately have registered about a 6.5 on the Failure-to-Communicate scale. If they wanted to let us in on the picture, Richter and Gutenberg and the rest of the seismologists goofed. They got so interested in what their calculations told them about what they didn't understand about the intrinsic nature of quakes that they seem to have forgotten us altogether. They created a scale where 0 is not the bottom and 10 is not the top, and where any of a half-dozen measures may provide very different Richter-type numbers. They correlated a "magnitude" scale to an "intensity" scale without taking into consideration how ordinary folks wouldn't understand the technical redefinition underlying these terms. And not to mention those goddamned logarithms. Simply put, logarithms aren't a humanized measure. It goes against all our counting-on-our-fingers mathematics to have a scale where 0 equals 1 and 9 equals 1,000,000,000. The only sensible way of conveying this is to metricize measure. You make 0 equal to a Richter and 1 to a decaRichter and 2 to a hectoRichter and 3 to a kiloRichter, etc., etc. Using this measure, you might end up in the unfamiliar realm of mega- and gigaRichters, but at least it would be clear that your measures are playing leapfrog with one another and not simply getting one number larger. But I don't know how you metricize a scale which smashes together two logarithms and thereby creates a situation where, energy release–wise, a 9 is upward of 20 trillion times bigger than a 0.

What should have happened, and what would probably happen today, is that some PR person should have taken a look at what Richter was devising and shouted: Whoa, fella, your scales are all very interesting to you high foreheads, but the ordinary people whom the quakes buffet and shake have a right to understand what you mean. So no dueling logarithms. No M_S, M_W, M_L, M_B, and their competing numbers. And, if you can't talk to the public within the framework of your original Richter scale, devise another one which makes sense to the average ten-year-old.

But that probably won't happen, because all scales with the name Richter attached to them no longer belong to us. They are owned by the seismologists. And if there is any lesson that can be gathered from the history of the world we now inhabit, it is that while science may give us ordinary folks wondrous technology and startling discoveries, scientists will never let us redefine a basic measure. That, as I am sure Charles Francis Richter would tell us from whatever numerical heaven he finally rests in, would be so impossible it is off the Richter scale.

Any measurement based on the classical Richter scale is, of course, completely artificial. Ergo: what we are looking at is how much the lines on the seismograph jumped when the earth moved. Anything that makes the earth move any amount could theoretically be turned into a Richter number.

In this vein, in 1969, University of Alaska geophysicist David Stone wrote a tongue-in-cheek letter to the scientific journal *Geotimes.* "If at any given moment all seven hundred million Chinese jumped off a two-meter-high stand, and assuming that their average weight is about fifty kilograms, then the energy released is approximately equal to an earthquake of magnitude 4.5 on the Richter scale," he suggested.

That quake would hurt only the Chinese, but if they organized their jumps to coincide with the natural period of the barely perceptible continuous ripple which continually sweeps the Earth's surface — that is, every 53 minutes — they could make the wave grow in a resonant effect, Stone theorized. And if they jumped on a part of that portion of the Earth where earthquakes and volcanoes occur — the so-called ring of fire which runs down the west coasts of North and South America, across the Pacific, and up through Indonesia, Japan, and China — they might trigger a great quake elsewhere. And if they were lucky, that elsewhere would be California.

The only antidote, warned Stone, would be if American politicians heeded the threat and organized their population to jump in between the heights of the waves and thus dampen them down. However, given the fact that there were many fewer Californians than Chinese, the mass crashing to earth would have to be from a considerably higher perch.

The only response in *Geotimes* was a letter from a Swedish scientist who pointed out what a creature of the Cold War Stone's letter was, and further argued that the science of the jumps required such a high degree of coordination that only countries with large populations and well-developed communications and monitoring

How many jumping Chinese does it take to cause an earth-quake?

systems could attempt it. He knew of only one such place where this existed in 1970: the United States.

However, *Time* magazine later picked up the story and featured it under the imaginative headline "The Great Leap Downward." They also added fifty million Chinese to the number of jumpers, and a minute to the wave time, but their most important contribution may have been to rouse others to similar calculations. One letter writer pointed out that "fifty million motorists simultaneously applying their brakes at a speed of 60 mph would impart 50 times as much energy to the earth's crust as David Stone's jumping Chinese. Careful coordination would focus the energy at any point on the earth. Unfortunately, the first Chinese jump would destroy our highways and prevent a retaliatory attack. Therefore, a pre-emptive strike should be made at once." Another pointed out that Stone's Richter computation reflected the simple physics of falling objects. That meant that to produce a 4.5 earthquake "the Chinese would have had to jump from 6½-ft platforms with *stiff knees.*" The most immediate result of that would have been not the destruction of California but "the ear-shattering scream from 750 million badly maimed Chinese."

However, with the end of the Cold War, it appeared to me that all this Richter silliness was going to pass into scientific oblivion. We need other measures of the scale-nobody-really-understands which speak to our peaceful era, I told myself. How big, I wondered, would a quake have to be to split the Earth in two?

It seemed a sensible comparison until numerous geophysicists subsequently told me that, in some sense, my question was a *non sequitur.* If the Earth were a solid, you might be able to split it with a quake, but, in fact, as U.S. seismologist Richard Stead pointed out, "the mantle exhibits viscous and plastic properties. If you attempted to 'split the Earth' you would mostly be dealing with stretching and flowing mantle, not fracture." Translation: The Earth's shell would crack, but its body wouldn't bust.

However, Stead soldiered on, imagine an Earth which was brittle throughout. If this were held in place, and if it were hit by a dense asteroid moving at a speed of 72,000 km (45,000 mi) an hour, then that Earth might shatter. The shock would produce a reading on the Richter scale of about 14; a shock that size would release fourteen-million times more energy than the largest quake the Earth has ever experienced.

That would, of course, be a tremor big enough to make all of us, and not just the Chinese, jump up. But it is not clear how many of us would be alive on the jump down.

The cooking-saurus

Cooking success is up to you!
If you'll take pains to measure true,
Use Standard cups and spoons all the way,
And then level them off — it'll always pay.
— *Betty Crocker's Picture Cookbook, 1950.*

"Are we going to measure, or are we going to cook?"
— New York Times *restaurant critic Mimi Sheraton's mother chiding when her daughter demanded recipe exactitude in compiling a cookbook.*

You begin to write a book on weighing and measuring and the world beats a path to your door. Well, not the world exactly, but friends and acquaintances *do* begin to advise you what measure mysteries they want resolved. Many have eccentric concerns — what the no-knock numbers mean in gasoline, how many leagues there are in a mile — but only one topic kept reappearing in my discussions.

Weighing and measuring in the kitchen. Because it was an area of which I knew little, naivety marked my reactions. Since everything we cook is weighed and measured, I would ask, what is the question? To which my culinary friends would respond with a litany of problems: scaling up and scaling down, correlating weights in European cookbooks to the volumetric measures favored in English recipes, and translating the cooking wisdom of the past — grandmother's scone recipes which said to "stir the dough until it is the right consistency to make scones" — into today's more precise terminology.

Oh, I said, how interesting (but really thought, how boring). Many of the issues seemed to be just innumeracy and North Americans' aversion to the metric system. How important could kitchen weighing and measuring problems really be, I told myself — and then I stumbled upon the Gulliver problem. In a way which none of my cooking friends could make clear, it showed me why measuring cup levels and oven temperature readings constitute a profound statement about what it means to be alive at the dog end of the twentieth century. The Gulliver problem, for those of you unfamiliar with *Gulliver's Travels*, was how the tiny race of Lilliputians could ensure that their giant captive, Lemuel Gulliver, was fed enough food to keep him alive.

The Lilliputians did not proceed in the most commonsensical way — by feeding Gulliver until he was full, and then toting up how much he had taken in. The decision to ignore this approach was in part a question of agricultural economics. The book tells us that some of the 15-cm (6-in) -high residents of Lilliput feared that the "Man-Mountain's" immense appetite would eventually produce a famine in their land. Therefore, they had to ensure that he got enough to eat, but not more than that.

Because Lilliput was a progressive and modern sort of place, the task of determining how much to feed Gulliver was given to the scientists. In Gulliver's own words, "his Majesty's Mathematicians, having taken the Height of my Body by the Help of a Quadrant, and finding it to exceed theirs in Proportion of Twelve to One, they concluded from the Similarity of their Bodies, that mine must contain at least 1,728 of theirs, and consequently would require as much Food as was necessary to support that Number of Lilliputians."

While they did not say exactly how they arrived at this figure, more than forty years ago Washington University zoologist Florence Moog calculated that the Lilliputian mathematicians produced this number by applying the well-known — to some — principle of dimensional analysis. This principle states that the weight of a body increases as a function of the cube of its height. Since Gulliver was just under 6 feet tall and the Lilliputians just under 6 inches tall, he was twelve times taller than them, so to arrive at his mass they multiplied 12 x 12 x 12 and arrived at the 1,728 figure.

In accordance with these measures, the Lilliputians tried to feed Gulliver an amount equal to the — lovely eighteenth-century word — victuals 1,728 of them would have consumed in a day — a figure, by the way, that meant slaughtering at least six Lilliputian cows and forty sheep daily, as well as moving wagonloads of other food and barrels of drink to him.

Unfortunately, but not uncharacteristically in this area, they were wrong, wrong, wrong in their scales, and consequently equally confused in their measures. The mistake was one which the best minds of the early eighteenth century did not appreciate. We now know that, if the Lilliputians' physiology had followed general scaling principles seen among mammals, they would have had a very different metabolism from Gulliver's. Their hearts would have beaten faster, their lungs consumed more oxygen, their intestines churned out waste matter more quickly than humans six times their size.

How much more? Professor Moog computed that, if the Lilliputians' metabolic rate approximated that of a 6-in mouse, then they would have eaten eight times as many calories per body weight as a full-sized man. This meant that, by applying the scale-up-from-us standard, they were giving Gulliver more to eat in one day than he could have normally consumed in a week.[1]

Because there is no complaint of overfeeding in the book, the skeptical Professor Moog suggests that a biologically naive Gulliver was having us on. The Lilliputians never existed at all except in his — or was that Jonathan Swift's — imagination.

However fun you find this, you might now well ask why Lilliputian measuring gaffes are of interest to thee and me? Well, while we may live in a world where species scale-ups are unnecessary, it is not at all clear that we humans have resolved the issue of how we are to measure cooking and eating. For example, if they followed today's nutritional bible, *Bowes and Church's Food Values of Portions Commonly Used*, modern Lilliputians would have to consider Gulliver's meal plans in terms of kilocalories; water; fat;

polyunsaturated fat; cholesterol; weight; protein; carbohydrate; saturated fatty acid; fiber; vitamins A, C, B_1, B_2, B_6, B_{12}, D, E, and K; choline; folacin; niacin; myoinositol; nitrite and nitrate; oxalic acid; phytic acid; phytosterol; purine; theobromine; tryamine; pantothenic acid; sodium; calcium; magnesium; zinc; manganese; potassium; iron; copper; phosphorus; chromium; cobalt; fluoride; iodine; selenium; tin; and biotin. And at the same time, in place of the unspecific measures of "barrels and carts," a modern Gulliver might find his meals prepared by chefs who worked within the scientific temple of cooking chemistry — the kind of world where, in 1950, *Betty Crocker's Picture Cookbook* could sternly advise its readers: "Measure as exactly as a druggist follows a doctor's prescription. Two minutes spent measuring carefully may save you hours of grief."[2] (The implication here is surely that the poor measurers are the potential poisoners.)

The questions in my mind are how we arrived at this mania for measurement in cooking, and what does it mean for the society we now find ourselves in? The most convenient tool with which to gain an appreciation of how measure-mad the modern cooking world is is that food-splattered relic of eating history, the cookbook. The historical differences in the presentation of recipes are not simply striking, but almost unfathomable. Following a tradition that could be traced back to the Greeks and Romans, a cookbook told you what went in and the order of their combination, but almost nothing more. To make spiced fish, *The Form of Cury* advised in 1378 "to take Lucys or Tenches and hack them small in gobbets and fry them in olive oil and seeth nym vinegar and the third part sugar and minced onions small and boil altogether and cast therein cloves, maces, quibbibs and serve forth." It seems as if almost everything was missing. How long do we cook, how small is "small," and how much seasoning is cast therein? I suppose what intrigues me most is that there is absolutely no mention of how many are to be fed. Over and over again in early cookbooks what seems to be assumed is that the words "too much" don't exist. The presupposition is that the cookbook users are preparing not for two, four, or eight, but for a multitude — for large families, their servants, their poor relations, and the beggars at the door. So, given the numbers of potential eaters, if you had the wherewithal it was always best to make as much as possible. As late as the 1860s in the United States, the average recipe seemed to assume that the average family numbered ten to twelve.[3]

Cooks groped their way to making cooking a measuring science in the way things usually happen in history: inconsistently. Initially what appeared was a sort of promiscuity of measure — you could be both precise and vague. In 1770, Englishwoman Elizabeth Raper wrote down a recipe not for food but for something else she made in her kitchen, lavender water: "Mix half an ounce of oil of lavender, 6 penniworth of ambergrease, mix them together and put to them a pint and a half of the best rectified spirits of wine, shake it well two or three times a day, let stand two or three months, then Philter it off."[4] No modern cookbook would mix measures in which cost and amount were equally regarded as worthy units. Although in these cold economic climes, where price and palate collide, the amount of hamburger that goes into a spaghetti sauce may well express a certain cost–measure elasticity.

As one moved into the nineteenth century, the metaphorical and the exact still inhabited the same cooking formula without any sense of impropriety. For example, William Kitchiner, in his snooty 1817 work *Apicius Redevivus, or The Cook's Oracle*, described a stuffing for roast pig wherein five ounces of stale bread crumbs, a handful of sage, some pepper, and "a bit of butter as big as an egg" are all included.[5] The same willingness to accept precision and imprecision was evident when time and temperature were at issue. Mrs. Raper was quite clear that, when making her rolls, you should allow the dough to rise for six hours, but then you "bake them in as quick an oven as possible" for a time that is not specified.

How to tell how hot was "quick" — the latter being a synonym for "very hot" — was a continual problem. While many of us may have a sense that measuring ingredients was a precarious business in the nineteenth century, it could always theoretically be more standardized than temperatures. Both the metric system and a standardized Imperial measure did exist, even if cooks seldom made use of either, not for the least reason that no standard cup or spoon was found on the market. However, before objective measures of oven temperatures existed, skin and paper were the only thermometers around. An 1888 American cookbook carried a folkloric bit of household physics: "To ascertain the right heat of an oven, put a piece of writing paper in it, and if it is a chocolate brown in five minutes it is the right heat for biscuits, muffins and small pastry. It is called a quick oven. If the paper is dark yellow it is the right heat for bread, pound cake, puddings and puff paste pies. When the paper is light yellow it is right for sponge cake."[6] If paper wasn't available, other books suggested, stale bread would also brown on cue.

It seemed to many cooks that imprecision was just what cooking was about. Mrs. Beeton, that eighth wonder of the nineteenth-century cooking world, summed it up. "If eggs had a uniform size, and flour always absorbed the same amount of liquid it would be possible to state precisely how many eggs or how much milk would sufficiently moisten a given quantity of flour. As matters stand indecisive terms and directions are unavoidable, occasionally something must be left to the discretion and common-sense of the worker."[7]

And a siege on common sense, critics argue, occurred when the passion for more exact measures suddenly gripped the cooking world. In North America, the change to "scientific measure" in cookbooks is generally associated with Fannie Merritt Farmer. In 1896, she published *The Boston Cooking School Cook Book*, now commonly called the *Fannie Farmer Cook Book.* The Boston Cooking School was an institution of which she was principal, and one thing that is striking about the book is what has been described as "her obsession with accuracy."[8] She was not the first to counsel that accurate measures were important. However, Fannie Farmer's zeal for correctness expressed a different science. Her book begins with one of the most famous sentences in modern cookery: "Food is anything which nourishes the body."[9] This sounds like a call for an omnivorous diet which will include slugs and monkey brains and mineral-rich mud. But that is not the case. For the next sentence defines the body for us in very non-slug, non–monkey-brain ways: "Thirteen elements enter into the composition of the body: oxygen, 62½%; carbon, 21½%; hydrogen 10%; nitrogen, 3%; calcium, phosphorus, potassium, sulphur, chlorine, sodium, magnesium, iron, and fluorine the remaining 3%. . . . Food is necessary for growth, repair, and energy; therefore the elements composing the body must be found in food."[10] The book goes on to discuss the body as a processor of the chemicals contained in food — what nutrient fuel can be found in what foodstuffs. But, as well, there is clearly a sense that food preparation had moved beyond simple taste. It was now a science. The wife and mother who prepared food had voyaged past counting smacking lips and quickly cleared plates as signs of her culinary success. She now had to weigh and measure in order to make sure that the body machine didn't break down. Thus, while attention was always given to Fannie Farmer's stern measure ethic wherein commandments appear in italics — "*a tablespoon is measured level. A teaspoon is measured level*" — it is the variety of scientific measurements which she encourages that is most striking. Table after table of calculations made by pioneering U.S. scientific nutritionist Wilbur Atwater were also included. There were various measures for "proteid [*sic*], fat, carbohydrates, mineral matter, water and refuse." Average daily requirements for people young and old were described in detail.

Clearly, what was happening was not so much a recalibration of how one cooked but a redefinition of food itself. Foodstuffs were seen as their constituent nutritive ingredients; they were a cache of calories, a plethora of protein, a fantasia of fat. Such quantification and analysis were Fannie Farmer's passions. Her students at the cooking school recall her going to the best Boston restaurants and, if she found a sauce whose composition baffled her, putting a few drops on a calling card and then carrying it away for further analysis.[11]

The cookbook she was most proud of writing was one for invalids and the sick, and one of her relatives described her as "a great executive, food detective, and gourmet, rather than a great cook herself."[12] From our vantage, Fannie Farmer looks most like a woman scientist — *manquée* — but, in fact, what clearly was also happening is that science and technology were challenging the old notions of what food was and how it should be used. Fannie Farmer was a vessel. The leveled measure was an expression of the caloric theory of body metabolism which Lavoisier and Laplace first put forward at the end of the eighteenth century and which Atwater was able to quantify with amazing accuracy by the end of the nineteenth.

Essentially, this view stated that, in energy, input and output had to be equal if an animal was to remain healthy. To determine the correct balance, the body was measured in ways which, in their single-minded reductionism, were a unique indicator of the arrival of a scientific society. Various chambers were built to test the amount of oxygen breathed in, carbon dioxide breathed out, and heat the body produced. Lavoisier and Laplace's initial experiment placed a guinea pig in a chamber surrounded by ice and measured the amount of ice which melted in a ten-hour period and the amount of carbon dioxide produced by the animal. This body heat is the root for the word "calorie." By 1896, Atwater and a colleague had published a summary of 2,306 metabolic experiments on people ranging in age from 18 months to 88 years, as well as on 1,362 animals. They followed on a previous tradition in which scientists turned themselves, their families, and their servants, and occasionally professional fasters, into experimental creatures.[13]

At the same time, a whole series of nineteenth-century experiments were finding that the correct healthy mixture of things was nonintuitive. Chickens fed polished rice in the Dutch East Indies developed a paralysis which was cured if they ate rice hulls.[14] The contributions of carbohydrates and fats to a sound diet were noted in 1830. The larger implication was clear. If the body was a heat-producing engine, then the cook was, in ways which had not been appreciated before, its engineer. Thus, while the standardized cups and measures made some difference in taste and thickness, these appear to be a kind of a side effect in Fannie Farmer's scientific universe. The cook measured accurately not only to make the same dish the same way time after time, but to make sure that the right amounts of the body's fuel were reaching the diner's corporeal motor. "Cookery is the art of preparing food for the nourishment of the body" is how the second chapter of the first edition of her cookbook begins.

And it is just here that a larger twentieth-century debate begins. The question is how much are we willing to "science-ize" those parts of our lives which are rooted in our pleasure centers? How much will we accept the word of systematic experimentalists if it is not coincident with the subjective measuring rod which is the pleasure of the tongue and the nose and the belly and the memory? Almost from the very beginning of the age of scientific cooking, critics of the demon weighers and measurers and reducers of food to chemicals have asked the subversive question: Isn't taste the highest arbiter of the cooking arts? Aren't the tongue and the stomach of greater significance than the leveled teaspoon? And aren't Fannie Farmer and her measurement fetishes the symbol of what is wrong with cooking in our time? "I'd rather eat what I'd rather. I don't want to eat what is good for me" are the prophetic words of one woman who was urged to accept a scientifically measured diet at the turn of the last century.[15] To this day, the art critics of the food world view Ms. Farmer as something like a fly in the soup of good eating. "The advent of Fannie Merritt Farmer was an historic watershed. Before her, women wrote of cooking with love; she made it a laboratory exercise. She embodied, if that is not too earthy a word, all of the major ills of 20th century culinary teaching," wrote a disgusted husband and wife team in the 1970s.[16]

And that criticism itself points to another large cultural issue embedded in the simple act of accurately weighing and measuring in the kitchen. For the appearance of the cook as food chemist means that ordinary people — and let us not be too coy here, for I mean largely ordinary women who were in one way or other excluded from the scientific world — were asked to incorporate into their daily lives the methodology of laboratory. Cooking a precise amount of ingredient at an exact temperature for a measured period of time attached you to the puissant engine of twentieth-century progress — SciTech World. Not only were the cooks required to be methodical, but, by extension, the eaters of these meals were asked to abandon the usual measures of goodness and try to convince themselves that, if something had the right amount of this or that, or was prepared in a precise and scientific fashion, then it must be good. In combining the ideas of scientific measures and measured meals, Fannie Farmer established a tension between science and humanity which has defined much of our century. How many of us today regularly incant a little mantra as we sit down to an enticing but theoretically unbalanced meal: I know this is bad for me, but it tastes sooooo good?

The question is, how has the effort to give eating and food preparation a scientific edge gone? The apparent answer is both positive and negative. Some measurement changes have been embraced so enthusiastically that tidal changes in food preparation have gone all but unnoticed. The instruments of weighing and measuring are now part of every kitchen armatorium. And while manufacturers in the United States in the 1920s were still producing a variety of "standard" cup sizes — some of which held 50 percent more than others[17] — today spoons and cups are all uniform, if one excuses the battle between metric and Imperial measure which still occurs in the United States.

Other changes have been equally dramatic. When the oven-temperature regulator arrived in the 1920s, it was so useful and so desirable that, as in many technological revolutions, it appears to have been adopted without anyone being conscious that several thousand years of assumed imprecision in cooking had been obliterated.

If the social record chronicled by women's magazines is any standard, there was a brief hiatus in which stoves came in with temperature regulators that experienced cooks weren't sure could be trusted. Women with experience with their own stoves "can tell by simply inserting a hand if the heat is right for bread or biscuit, loaf cake, apple pie or cream puffs," opined Minna Denton in *Ladies' Home Journal* in July 1922, adding, "it would be folly to expect a mechanical device to do away entirely with the necessity for using one's judgement, but it is equally foolish not to avail oneself of every help experimental science has to offer."[18] However, in 1923, *Good Housekeeping* magazine was encouraging its readers to accept the fact that temperature control was now an intrinsic feature of cookery. "Do you cook by temperature?" it asked its readers in that hectoringly pedagogical style which remains a feature of similar magazines to this day. "If not, do it now, and safeguard your cookery methods."[19]

The only hiccup in the adaptation of the controlled-temperature oven, and one which anyone who has struggled with the maze which computer incompatibility has experienced, is that no one informed the stove manufacturers what standard they should hew to. The result was a kind of anarchy of temperature languages. Some stoves had gauges which told you actual temperatures.

Others used numbers 1 through 8 or 9. Still others used letter codes C through G, or D through J, as gauge stops. This meant, among other things, that the temperatures represented by numbers or letters on one manufacturer's stove weren't necessarily the same as those on another's.

Consequently, by the 1930s, English cookbooks were providing technological translation tables for cooking by temperature. Fahrenheit, Celsius, and Reaumur scales were correlated with the various number and letter scales, and then translated into the old measures of quick or slow.[20]

In a more general sense, modern cookbooks proclaim that exact measurements have intellectually swept away the imprecision of previous centuries. If the early cookbooks had no measures, their descendants today seem to calibrate everything. *The Larousse Traditional French Cooking*, published in 1989, breaks down recipes into both preparation time and cooking time so that the cook knows the length of time from start to the table. It gives exact measures for ingredients in both Imperial and metric units. It is a little laissez-faire when it comes to temperature, speaking in terms of "hot ovens" or "warm ovens." However, a table at the front of the book translates these measures into degrees Celsius and Fahrenheit, and gas-oven mark scales. It advises that, unless otherwise indicated, the recipes serve eight. "Health" cookbooks will tell you exact caloric and cholesterol levels in the foods you prepare.

However, the triumph of measurement may not be as far-reaching as it seems. It is not clear to me that people actually follow slavishly the recipes they are presented with. In part, I have derived this impression from a whole cadre of food critics and restaurateurs who vehemently maintain that cooking — at least, good cooking — is an art and more than a science. As chef and cookbook writer X.M. Boulestin haughtily put it, "The most dangerous person in the kitchen is one who goes rigidly by weights and measurements and thermometers and scales."[21]

Equally problematic is the question of how willing people are to treat eating as a kind of measured, scientific nourishment of the body. One continual problem has been that the science in this area has been "real science," by which I mean something that not only bumpily evolves but blithely changes its mind as evidence of one sort or another accumulates. If there is anything outsiders find continually irritating about the scientific method, it is that there is no bad conscience about mistake or miscalculation. If you alter your whole eating life based on a study which is overthrown a year later, nobody is going to beg forgiveness for the troubles their errors have caused you. And for those who initially wanted to believe that obsessive calibration was a kind of swindle on the senses, hardly had Atwater concluded his studies when strong evidence of errors in his measures began to appear. He seriously overestimated — by a factor of 2 or 3 — the amount of protein which a person needed for a "balanced diet," not to mention ignoring entirely the contribution of not-yet-discovered vitamins and amino acids. But what better indicates our twentieth-century confusion about what we should eat than the fact that no fewer than six revisions of the U.S. Department of Agriculture's food guide to healthy eating have been made since it first appeared in 1916. The number of food groups have bounced around from 5 to 12 to 7 to 4, back to 5, and up again, to 6. The first USDA food guide included "flesh foods, starch foods, fat foods, watery fruits and vegetables, and sweets."[22] Over the years, sometimes tomatoes and citrus fruits have been their own category, as well as fats, sweets, and alcohol, and during the 1940s there was a butter-fortified margarine group. Even today there is no agreement as to which is right or better or more in tune with how people actually approach their tables. "Even those who assert that food guides should teach why foods are eaten disagree whether this means emphasizing foods or nutrients," comments a group who has written a social history of U.S. food guides.[23]

Worse still, in North America at least, overeating has overtaken hunger as the largest dietary villain, and food guides have taken up the cudgel, telling people that they *shouldn't* eat what they *are* eating. The swings in nutritional advice in food guides are so wild and so regular that in 1991 a group of radical vegetarian doctors gathered large media interest with a new food-group chart which recommended that the healthiest diet had no animal product of any kind in it.

At the same time, there have been rises and falls in the recommended amount of calories, fats, proteins, carbohydrates — nearly everything that a person is supposed to eat. Between 1948 and 1974, no fewer than eight different estimates were made by a variety of food organizations around the world about the amount of protein that a child under one year of age requires. In the United States alone, the number of grams per kilogram of body weight that should come from protein was reduced from 3.3 to 1.35.[24] In Europe, in 1990, recommended daily allowances ranged from 81 grams in France to 54 in Spain.[25]

One hundred years after Fannie Farmer, we still don't seem to be able to agree about what the kitchen scientist should put on the table and how that relates to what his or her family likes to eat. And there may be no better indication of the confusion than the complete capitulation of Fannie Farmer's cookbook itself. In later editions, it gave up its romance with food "science" and replaced Fannie Farmer's bold, if cold, original declarations with smarmy, unscientific rhetoric. In place of "food is anything which nourishes the body," the twelfth edition of the cookbook, published in 1990, proclaimed: "Every meal should be a small celebration. If you acknowledge so joyous a fact of life, the pride you take in your efforts in the kitchen won't be confined to company occasions." Does anything suggest more about our uncertainties about the place of weighing and measuring than having *Fannie Farmer* burlesque its scientific beginnings? All of which brings me to what is, in effect, a second beginning to this chapter. Having discovered what the history of food measurement seemed to be saying, I wondered how its truths were applied by cooking's practitioners. And so I spent an evening watching one of the people who had encouraged me to write about kitchen measurement — my incredibly energetic and good-hearted colleague Stevie Cameron — wrestle with her own version of the Gulliver problem. In a downtown church she was busying herself and others preparing a multicourse meal for a hundred or so of the hungry and homeless men and women coming in from cold, November streets. Her Gulliver problem was how to combine all the differing tastes and nutritional needs into one meal. As she stirred, and tasted, and gave orders to a dozen or so helpers, I tried to concentrate on only one thing: How does she measure? It was most surprising. First, like the old cookbooks, there was a vagueness about how much was being made. No one knew how many people might show up, as anywhere from 100 to 130 people had appeared in the past. So the amount that she and her dozen or so helpers made was closer to "a lot" than any fixed measure. Three ovens were used to cook the meals. And while each came equipped with a temperature-measuring gauge, one was known to be a really hot oven. When it read 350 to 375°F, the food preparers had learned it was closer to 500°F. And that meant you had to be doing an *ad hoc* recalibration of how long you cooked potatoes, for instance, taking into consideration a burn factor and remembering what you had put into what oven. In all, it was redolent of the nineteenth century. As well, I never saw Stevie measure anything except the amount of cous-cous. How, I asked, did she know, for example, how much béchamel sauce was needed to cover a pan full of cauliflower? "I am a very experienced cook," she replied, as one might to a kind of willfully ignorant child, "and because I am experienced I know I needed three liters of béchamel, and I know how much I put in a recipe at home and I tripled that." It was exactly the sort of thing that you can imagine the Roman chef telling a dunce apprentice when he or she asked for some kind of measure precision in a given recipe.

As I left the church for the evening after watching the people attacking their meals with a silent intensity which I recognized from those times in which hunger had taken over my entire psyche, I wasn't sure how to sum up what I had seen and what I had learned. I think there is a natural limit to measuring both foodstuff and what is in it, because most people don't like turning existence into a laboratory life. Compulsive measurement limits our sense of a subjective self. It is a reason so few of us become scientists. It is the reason so many of us distrust an approach that seems more precise than anything in our ordinary life. If there is too much accuracy, the result becomes not *our* cake or *our* soufflé, or even *our* meal, but the method's product. In response to the question "Did you like it?" who wants to hear a satisfied guest respond, "Yes, it was lovely. I see that you followed each measurement jot and tittle like a good little cuisine compulsive."

In another sense I am tempted to suggest that insofar as nutritionists have failed to arrive at a measure of the proper diet, they have undermined society's larger faith in science in general. A bright, doubting child may well ask: If the best minds can't tell us from decade to decade something simple — what is right and proper to eat — how can we believe them when they say they know something almost unimaginably complex — like how the universe came into being?

Ultimately, I suppose I must conclude with the riddle of modern times. *Homo sapiens sapiens* are, despite the sheen of a technological veneer, very much the creatures they have always been: the one species on the planet which thrives on complex contradiction. We are, in measurement, as in eating habits, omnivores.

Le porko-metric

For almost as long as I have written about science I have stumbled when trying to quantify the economics of political decision making. How do you compare the incomparable, and measure the immeasurable? Should a billion dollars go to AIDS research, or would the money be better spent by sending a robot to Mars?

I had almost come to the conclusion that no measure was possible, and then one day I found myself half-dozing through a symposium on a Canadian effort to establish a conservation department at the Museum of Carthage in Tunisia. That museum had started out as a kind of scholarly holding tank for the artifacts which a French religious order had unearthed during their excavations of the ancient city. These had not been well preserved after the museum became a property of the state. Consequently, Dr. Ursula Franklin, a professor of metallurgy at the University of Toronto, and several other scientists had spent three years showing the Tunisians how to restore and preserve the 11,000 artifacts that bespoke Dido and Hannibal and a vengeful Rome. The preservation was a noble task, surely, and one whose significance defied measure. But a measure issue did arise when some doubting voice in the audience asked the ugly question: How much had this beneficence cost Canada?

"It cost $500,000." Pause for effect. "But that is only one-quarter the price of a single cruise missile," said Franklin, who is that wonderful contradiction in terms known as a militant pacifist.

It was a singularly effective riposte. For I believe that most of the audience reacted initially as I did: only a quarter of a cruise missile? Why, who could even bother to think for a blink about such a minuscule expense? The Carthage project, expressed in cruise-missile metric, turned into a pittance, into what W.C. Fields has characterized as a "mere bagatelle." It seemed only appropriate that she then used this comparison to launch an appeal for her government to turn some of the late twentieth century's laser-guided sword money into the plowshares of museum preservation.

But the next day, as I was going over my notes from the presentation, I realized something almost perverse had happened. Dr. Franklin had militantly clobbered us with a context. The force of her argument was almost entirely dependent on the war-mongering yardstick she had chosen to measure the Carthage museum project against. Someone else, someone less of a militant pacifist and more of an economic rationalist, might have taken her $500,000 bagatelle and made it huge and wasteful. He or she might have suggested it was the cost of inoculating 50,000 Tunisian children against six common childhood diseases. It was dollars enough to provide 4,000 Somali families with enough seed and tools to recover from the drought; it was the hope for a Newfoundland outport enduring a fishless winter following a fishless summer. It was . . . But you understand what I am talking about. And for a second I admit I experienced a sense of despoliation, that gray, greasy sensation which oozes into your stomach when you realize that a politician is lying to you and loving every minute of the lie. But then I pulled back and understood that Franklin wasn't corrupt or mendacious. Rather, she had shown that, in an age in which we have measured everything from the time it takes for an eye to blink to the bursting point of a condom, how creative people can be in places where no standard measure exists. In a vacuum people are free to invent their own scales and measures and try them out on the rest of the world.

Moreover, the question which had so long troubled me — How do you compare one government "good" with another? — appeared a particularly active area of measure invention. Further research uncovered that satirist P.J. O'Rourke, who has a genius in this arena, has computed that what the United States spent to bail out of the savings and loans companies — $500 billion — is equal to the amount of money it would cost him to make ten round-trip taxi rides from Manhattan to Uranus at the usual rates charged by a delighted New York City cabbie. And yes, there would even be enough left over for a tip. Princeton physicist Freeman Dyson has described how one of his colleagues explained to the U.S. National Aeronautics and Space Administration that he could provide reflector units which Apollo astronauts would place on the moon for $5,000. NASA demurred, saying the reflectors that would become part of a laser system which measured the motion of the moon had to be put out to tender. The tender produced reflectors at a cost of $3 million a unit and thus established a *de facto* standard of NASA waste. Anything a private person can do, the space program can do 600 times more expensively.

While these had their allure I soon came to believe that a close examination of the features which make various measures appealing would allow me to arrive at the first principles of a truly international measure of worth and waste: a porkometric. It would be my contribution toward making the world a more rational place. But first, since the new word comes from the American political slang term "pork barrel," a little linguistic history is in order. The term "pork barrel" comes from a practice associated with slavery. On special occasions, plantation owners would dispense extra rations of salt pork, which were kept in a barrel. By the 1870s, the regular stampede of Congressmen trying to slip local appropriation items for their constituencies into larger spending bills was being likened to the rush of slaves running toward their rare treat. The term today means something like "government spending with little purpose other than to put people to work."

How does one quantify a pork barrel? Dr. Franklin has shown us the virtue of framing your pet project so as to make it pale in the shadow of something else whose expense and uselessness may be deemed astronomical. I take this to be the basis of Strauss's First Law of the Porkometric: An Enormous Measure Is A Good Measure Because It Makes Everything Else Look Small. Military spending in nations officially at peace seems rife for these kinds of comparisons. In the United States, immoral measures range from the cost of the cold war ($10 trillion) to the national debt ($4 trillion), to the cost of the Strategic Defense Initiative ($1 trillion). A U.S. pacifist organization in 1991 compiled a list of what one could do with 5 percent of the world's annual $1-trillion military budget and claimed that one could eliminate starvation and malnutrition, prevent global warming, stabilize the world's population, immunize all of the world's children, and still have billions left over.

When he is not writing novels, describing the universe, or analyzing the possibilities of a nuclear winter, Carl Sagan is one of the masters of big-made-little porkometric scaling. In a 1985 analysis of the Strategic Defense Initiative (SDI), he played with the $1 trillion SDI was supposed to have cost. To help us understand how porkometrically bad the SDI $1 trillion was, he translated its cost into useful and worthwhile good expenditures. It could have been used to retire "more than half of the national debt" and pay off the money the Third World owed the West twofold. However, then he casually slipped in his special pleading. "With a fraction of a trillion dollars we could establish permanent human settlement on Mars," he said. What he conveniently left out is that this "fraction" amounted to somewhere between $100 billion and $500 billion — a number which the opponents of the On To Mars initiative have translated into an awful lot of, in their eyes, more worthwhile roads, bridges, and school lunches. Sagan could do this because that was the genius of the SDI porkometric measure — it allowed him to make just such a comparison in a way which made $500 billion seem small and worthy.

As suits a smaller economy, the Canadian attempts to find a measure of big and immoral have been less grandiose but clearly have revealed a corollary to the first law: If Your Measure Kills People, It Will Make My Request Look Smaller.

A few years ago the Canadian government of the day floated a proposal to purchase nuclear-powered submarines for a cost of $5.6 billion. From the instant that they were suggested, the nuclear-powered subs (NUPOS) became the *de facto* metric measurement of Canadian government waste. What was also clear is that a system based on a NUPOS obeyed the first law. The money to be spent was far greater than any normal person could comprehend in his or her own budget. And — and this was significant — there was a special moral stink to the project. It wasn't just submarines, it was *nuclear* submarines. Only an extremely primitive mental extension was required to turn the *nuclear* subs into *nuclear* bombs, which, as we all knew, spelled the end of mankind. Thus all references using a NUPOS-based scale carried this hidden message: Not only is my project grotesquely cheaper than a nuclear submarine, but it won't kill anyone.

All our attention until now has been focused on a comparative measure, but this is only part of the story. Sometimes you want to make your project look smaller and bigger at the same time. The proposed international space station exemplifies Strauss's Second Law of the Porkometric. It states: Sometimes The Smaller You Can Divide The Parts, The Bigger You Can Make The Whole. The idea of the international space station has been attacked since its inception for being both too costly — at one time over $30 billion to build and an extra $100 billion to operate over its lifetime — and of not much use either to science or to industry. However, every year when it comes time to kill the project the U.S. Congress has shown a surprising reluctance to lop off the station's head. Why? Wisconsin Democrat David Obey has pointed out that "NASA has successfully salted this program so that they've got contracts going on in 218 Congressional Districts." That is one more than half the votes needed for a majority in the Congress and clearly is something which changes the space station from one big wasteful project into 218 smaller, politically desirable ones.

This division into units of smaller benefit is a defining characteristic of the pork barrel. The principle is that, in order to sell a project, you must change it from one large, fiscal ham hock into sausages which agglomerate a host of smaller but somehow more valuable pieces of pork.

Conversely, sometimes a big project needs to look bigger than it is. This leads to Strauss's Third Law of the Porkometric: A True Porkometric Scale Will Have Been Arrived At When All Costs Have Been Translated Into Much, Much Larger Benefits. As an example, people in British Columbia have been pushing for the construction of a particle accelerator which goes by the name of the Kaon Factory. These accelerators smash atomic particles into one another in order to see if there are smaller particles inside. In order to justify its bulging, nearly $2-billion price tag, the Kaon project leaders hired an outside team of consultants to compute the amount of spinoffs the accelerator would generate. They concluded that the yearly $60-million operating expense would eventually spawn $243 million a year in ancillary revenues in fields as far-ranging as remote terminal units for refineries and medical control systems. Thus, what to all the world seemed like a huge expense, by their calculations turned into a cash cow constructed out of the byproducts of smashed-up particles.

The Canadian federal government looked at British Columbia's measure of expanded good, something which I will call a Kaon, and found it to be somewhat suspect from its inception. Not only was there not the slightest inducement for the outside consultants to be pessimistic about their projections, but a later study found the scenario of Kaon as cash cow was "explicitly optimistic . . . [with] undiscounted, gross revenue estimates [that] are therefore inappropriate for inclusion in a benefit–cost framework." And worse, while the analysis was able to find economic benefits from the program, it was about the same good you would have gotten from any similarly sized and similarly expensive project.

Another tactic that is often used is to translate a proposal from something strictly monetary into time or space measures. This is Strauss's Fourth Law of the Porkametric: Buddy, Can You Spare

A Million? This is such a common ploy that its cleverness is overlooked. You don't say a project will cost $250 million, you say that it will cost only $1 for every American. Who is such a piker as to begrudge one dollar? Or if the figure is a lot bigger than that, you suggest that for a mere 50 cents a day — a mere $45.6 billion yearly if every American contributes — you can feed the starving children of the world.

Another way of accomplishing the same thing is to turn money into distance, or as the corollary to the Fourth Law states: When You Don't Like What Something Costs, Make It Go To The Moon. The best example of this corollary belongs to a *U.S. News and World Report* article which was able to translate federal spending into something an ordinary person would gag over. In 1984, they computed how much each president spent per day of his term of office. Wise in the ways of simplification, they avoided such nasty complicating factors as changes in the physical size of the country, explosions in the number of people living in it, and inflation. Being as simple-minded as possible, they were able to show that the alleged anti–big government tax-cutter Ronald Reagan was spending nearly $2.6 billion a day. In comparison, George Washington, he of true-blue parsimony, had overseen a daily dole-out of only $20,548.

In case the reader wasn't suitably impressed with a few billion dollars a day flowing down the government spigot, the magazine translated this into bigger-seeming numbers. If "laid end to end, 2.6 billion $1 bills would stretch 251,103 miles — 12,246 miles farther than the mean distance between the earth and the moon." Then some bright light must have realized that their yardstick — Earth to Moon — was beyond the average nonastronaut's ken. Got to make it homier. "Piled flat," they continued "one on top of the other, the 2.6 billion $1 dollar bills would stack 176 miles high." Oops, that may have diminished it a bit too much. The distance you travel in just over three hours going at the legal speed limit is too small a number. Therefore the wise heads at *U.S. News* compared their distance/dollars against a smaller measure to make it look bigger. The 176 miles translated into "745 times as tall as New York's Empire State Building." Of course one would be a spoilsport to point out how a different approach could have shrunk *U.S. News's* pork. If $100 bills were used, the deficit was only 7.4 times as high as the Empire State Building. And if one had compared the deficit to something truly huge — say, a light-year — it turns out that end-to-end U.S. deficit is only about 0.000000427 of that. A piddle. A mere bagatelle.

All of these principles are confusing, and thus it has become clear to me that we need to establish a standard waste-measuring unit. I have thought long and hard on this, recalling the terrible travails the French faced in establishing the exact length of the meter and bearing in mind that my porkometer should be something universal, elastic, and funny. It should be something which will remain when the civilizations using it have gone broke. It should have the capacity to be both big and small, to turn a cost into a benefit, to be exaggerated beyond belief and shrunken into near nothing. It should be divisible into as many parts as a government patronage policy can devise. I considered Ursula Franklin's cruise missile until a reading of various documents indicated that its cost ranged from $1.3 million to $3 million. Besides, it comes apart only when exploded.

I have thought over and over, and then I recalled the Rhinoceros Party of Canada. As one of their prankish election planks, this grouping of poets and jokesters and don't-give-a-damns once proposed that the unemployed be hired to level the Rocky Mountains. It was more than a make-work proposal; it was an unmake-planet proposal. A gargantuan measure, but of course computing how you would dismantle mountains is fraught with, as the sociologists say, methodological problems. Can you use dynamite? Do you build a railroad to cart away the debris, and is that a spin-off benefit? And would Rocky Mountain Porkometric be appealing to everyone, would porkometric-using environmentalists find themselves soaked in spittle by fractious brothers-in-arms who saw RMP implicitly condoning habitat destruction?

No, we need to remove Rocky Mountain Porkometric from an actual object of measure. What we need to imagine is a string of mountains passing through every nation on Earth. Each pretend mountain is a perfect pyramid 7,620 m (25,000 ft) high and 58,100,000 m^2 (625,000,000 square ft) at its base. Thus we have about 147,000,000,000 m^3 (5,208,000,000,000 cubic ft) of — oh, did I forget to mention that my mountains are giant radioactive dung heaps. Not any dung, but purée of pure pig manure and, as such, easy to shift. Let us assume a single person — an imaginary individual with a dung-shoveling capacity somewhere between the average man and the average women — can remove 1,180 cm^3 (72 cubic in) a shovelful. If we figure that between meals, and rests, and gossip, and trips to relieve themselves they can shift 10 shovelfuls a minute over the course of a day, we end up with 17 m^3 (600 cubic ft). If we assume that the shovelers work an eight-hour shift all year long, then three people working in tandem would take 8,680,000,000 days, or something approaching 24 million years, to move a dung mountain. If each person was paid $2.30 U.S. a day — a kind of world standard-of-living average — then it would cost in 1994 dollars roughly $60 trillion to shift the pig purée from one place to another. After which the new mountain would be moved somewhere else.

I have given this perpetually moving project a name. Each mountain will constitute one PigDung, and as such will become the standard measure of the pork barrel. You can make its costs and moving times bigger by making the shovels smaller and, in so doing, employ more people. You can see its storage as either an environmental benefit or a loss of good natural fertilizer. You can lobby to have a PigDung-sized project moved to your jurisdiction, or you can argue before the legislature that your daycare center must be worth at least a fraction of a PigDung make-work scheme.

You can also show that people are still willing to spend forty times more on killing one another than on the peaceable act of manure shifting. How much better, you could argue, that there be 10,000 PigDungs than even one nuclear weapon. *Au contraire,* what kind of corruption went into diverting our tax dollars toward the endless reshuffling of the dung heap in the first place? And what better condemnation of our or your or anyone's political system than the leveling and raising of Mount Pig Poo?

Yes, the PigDung satisfies the needs of a Porkometric. If only we can find politicians honest enough to demand a standard measure of government spending. But I fear I have fallen into another contradiction in terms: honest politician.

The ozone-saurus

My daughter's science teacher recently corralled me during a parent–teacher night at her school. Would I speak to the students? I instantly responded yes, and then mentally backpedaled as I realized that it is not the dream of most grade nine students to have their father spew his views before their class. But the teacher missed the pause and immediately asked what I would speak about. Well, I replied, I think that the most important thing to teach children in an environmentally conscious age is alternative views of nature. They must be shown how our interpretation of natural systems is often completely dependent not on what is there but on what kind of a box we draw around the data. And if they are going to be smarter than their parents, then schoolchildren must think subversively about accepted wisdoms concerning natural systems.

Now it was the teacher's turn to pause. He thought, I suppose, that if I believed there was anything that fourteen-year-olds didn't know about thinking subversively, then I was probably naive beyond reason. What sort of examples would you use? he finally asked. Ozone and ultraviolet radiation, I replied, and in particular how and what went into making the measuring system for UV radiation that newspapers and television now use to advise people of when it is safer to go out in the sun and when it is more dangerous.

Oh, that *is* interesting, he said, adding: I remember when I was in California ten years or so ago, you would lie on the beach and the local radio stations would announce every fifteen minutes that it was time to turn over. They didn't do it because they wanted to warn you that the sun was dangerous, they did it to ensure you got an even tan.

Boxes around natural data, I said, but that is nothing compared with what happened in making Canada's and the United States' UV index. And since I could see that he wasn't prepared to hear my entire spiel there, I decided to make the issue the subject of this essay, whose title might be: "Everyone Who Has a Yardstick Measures His or Her Universe in Yards."

Before I begin I should warn you readers what I am *not* doing. There is no suggestion within this essay that human industrial emissions have not contributed to the depletion of the Earth's ozone. A skeptical Stephen Strauss makes no defense of a life of endless sunbathing and suntanning. And anyone who doesn't think that the development of certain kinds of cancers can be influenced by the time we spend in the sun should look for support elsewhere. I issue this advisory because in my experience even the gentlest debunking of environmental science is a little like trying to milk a porcupine. No matter how careful you are, you always end up with quills sticking into you.

This is an essay about measuring — in particular, it is a reflection on three different measuring groups with different measuring instruments: politicians, atmospheric scientists, and medical researchers. Each, as you will see, drew his or her own box around the data at hand and came up with a differing image of nature.

First, the politicians. In February 1992, the U.S. National Aeronautics and Space Administration (NASA) held a press conference at which they announced they were scared of what they had seen in some preliminary satellite and plane data. The data showed that there was some significant probability that a hole would appear in the ozone layer over the Arctic, and that 20 to 30 percent of the ozone in northern latitudes would temporarily be lost over areas spreading as far south as New England and France. This would also mean that significantly more solar radiation would reach the Earth because the three-oxygen ozone molecule was known to block something between 90 and 99 percent of ultraviolet radiation in the wavelength that most affected human skin. The "Arctic Hole" prediction seized everyone's attention because just such a "hole" (I'll explain later why "hole" is a bad word) had appeared in the Antarctic seven years earlier. (For the record, NASA's prediction was predicated on weather conditions remaining extremely cold in the Arctic atmosphere. Ironically, subsequent atmospheric readings taken on or near the day of the press conference would show that the Arctic weather had warmed and that no hole, or even a dramatic decline, in the ozone was going to occur.)

The politicians in countries that might be affected reacted to the announcement as if radiation were already pouring down on every man, woman, and billy goat. Senator Al Gore, now Vice-President of the United States, took the tack that the preliminary data meant that one must push for a faster reduction in the chemicals which were known to destroy ozone. "Now that there's a prospect of a hole over Kennebunkport [then-president George Bush's summer residence] perhaps Bush will comply with the law," he thundered. In Denmark, a ministry of the environment spokesman urged his countrymen not to panic, but instead to wear hats and sunscreens. And in Canada, Jean Charest, then minister of the environment, had the strongest reaction. The man whom journalists have described as the most instinctive politician of his generation felt comfortable telling reporters that, in this time of crisis, Canadian children should be kept inside. "It is a lot more prudent to stay out of the sun." Given the fact that nothing had yet happened, all the politicians' responses were, let us say, premature.

To many scientists this was the worst kind of ignorant politico knee-jerkism. Lucien Froidevaux, one of the NASA scientists whose findings had so stirred Mr. Charest, told me, "I don't think there is any evidence of a real huge depletion in the ozone. . . . I wouldn't worry about my kids if I were in Canada, myself." But then again he wasn't a politician who is measured in some way by his or her ability to react to situations and to appear to be leading a response. Mr. Charest, after fending off claims that he was crying wolf, set in motion a plan to translate ozone loss into a number that would tell people just how dangerous it was to be outside on any given day. Officially, this daily scale was supposed to be under development, as an element in something called the Green Plan.

(This was an environmental umbrella that had been put forward two years earlier.)

In truth, next to nothing had been done about a UV index before the NASA alarm. And so, in the aftermath of the minister's jerking knee, the atmospheric physicists got together with the skin doctors and with public-relations people to try to come up with a scale that would translate changes in the amount of ozone into an understandable measure. And it is here that I would tell Anna's class that the very different ways in which varying sciences look at and measure the natural world manifested themselves. Everybody, it seems, has his or her own yard from which to make a UV yardstick.

To begin with, there was a whole historical measure of ozone that was driven into being by nothing so much as scientific curiosity about what our atmosphere is composed of. It begins with smell and electricity. Ozone's distinctive funny odor following its creation by lightning had been noted as far back as Homer, but in 1789 a Frenchman suggested something in the atmosphere was absorbing most of the ultraviolet region of light from the Sun. The next year an Englishman did some laboratory measurements of ozone absorptive capacity and suggested that the ozone molecule was the material responsible for the lack of UV reaching the Earth. In 1912, two Frenchmen, Fabry and Buisson, made further measurements and concluded that a vertical thickness of atmosphere contained very little — 0.5 cm (0.2 in) — ozone. They later suggested that a layer of ozone had been formed by solar radiation's interaction with oxygen and was mainly situated at a height of about 40 km (25 mi).

Subsequently, just such a layer was found, with the greatest thickness being at about 25 km (15 mi) from the Earth's surface. There is a semantic problem here. Even at its thickest there was no more than 1 ozone molecule for every 100,000 molecules of ordinary oxygen and nitrogen. So, at its most intense concentration, ozone is more like a flavor than a layer. Over time the study of the flavor/layer turned up a variety of measurement artifacts about atmospheric ozone. It was twice as thick at the poles as at the equator — at least, during late winter and early spring. There were seasonal variations that made outside equatorial regions see ozone levels go down in summer and early fall by as much as 25 percent and then rebound. And there were daily weather changes of as much as 10 percent. The thinness of the ozone meant that no true hole in the atmosphere would appear if ozone went down — the "holiness" scientists saw was an artifact of satellite imaging technology.

As befits a science, the study of ozone had its own measurement, the Dobson Unit. One hundred Dobson Units equals 1 millimeter of pure ozone at sea level.

This was not the total extent of human interest in the effect of ozone depletion. Medical scientists had spent the better part of a century trying to measure and quantify the effect of sunlight on humans. What they were increasingly concerned about was the effect on living matter of the small amounts — somewhere between 1 and 10 percent — of the most potent skin-affecting ultraviolet radiation that came through the ozone. I don't want you to strangle intellectually on the physics of light, but for what follows it is important to keep in mind that UV comes in four wave bands: Vacuum UV, which is less than 200 nanometers in length; UV-C, which is 200 to 280; UV-B, which is 280 to 315 or 320; and UV-A, which is 315 or 320 to 400 nanometers. The first two do not reach the Earth and so are not of concern in this discussion, but the last two bands do make it wholly or partially through, and so create the pressing public-health reason behind the UV index.

The formal study of the relationship between health and sunlight has had its paradoxes. Some diseases — smallpox, for one — exhibited greater scarring if a person was in the light. Others, such as skin tuberculosis, healed more quickly and with less scarring with exposure to sun. Scientists later discovered that a certain amount of ultraviolet radiation stimulated the body to produce vitamin D. From a measurement point of view, the question of whether there was a way of quantifying the relationship between sun and skin in physical terms arose. In 1927, two Germans came up with just such a measure, a Minimal Erythema Dose, or MED. Roughly speaking, a MED is the amount of UV dosage that it takes for the first flush of red to appear on the skin after an untanned person goes out into the sun.

Now if a MED seems like a straightforward Dobson Unit, you know nothing of biology. Biology is a measurement trickster, and this is something which we and Anna's class must never forget. While initial burns could eventually be differentiated in terms of just what wavelengths of ultraviolet produced the quickest reactions — this is the reason we know UV-B is so potent — there was no 100-percent accuracy in determining anyone's burning time. Different skin types react in different ways. A Nordic Caucasian whose skin never tans will burn at a tenth of the ultraviolet stimulation needed to burn an Australian Aborigine. And each time we are out in the sun changes our next burn time, as our skin thickens to deal with the UV-B. If you start to turn red, the next time it will take a MED that is 20 percent higher to produce the same blush.[1]

More importantly, MED measures were done in a kind of pure science vacuum. What was true in a laboratory, with a calibrated sunlamp that gave off an exact amount of UV energy per minute, was one thing. But in the truer meeting of people and sunlight, confusion was everywhere. Consider the various confounding factors. White sand may reflect back as much as 30 percent of UV-B — one reason you get a quicker burn on the beach than in your backyard. And white snow may effectively reradiate all the UV that strikes it, a reason many spring skiers look parboiled. The angled, less intense sun which shines at eight o'clock in the morning takes 46 minutes to start to redden the skin of a person who burns before tanning, but only 11 minutes at noon. And then there are the clouds. High, streaky cirrus clouds block only about 16 percent of the UV-B, while the fleecy nimbostratus filters out upward of 80 percent of the rays.[2] And there is geography. Because the ozone layer naturally thins as you head south, an 8 percent drop in ozone over Toronto, Canada, means the amount of UV-B that reaches the ground equals that of Washington, D.C., where the ozone layer is naturally thinner. And while light ground fog may not block much ultraviolet radiation, ozone-rich smog does. Measurements in Mexico City, dirty-air capital of North America, show that smog alone can cut ultraviolet radiation levels in half.

What I am saying in all of this is that people trying to work out their own MED are likely to find themselves faced with labyrinthine calculations. The difficulty of the problem may explain why surveys taken of sunbathers show that they continually underestimate their burn times. But all of this overlooks a more important issue that the Canadian group had to deal with. The MED's concern with reddening skin is an abstract measure; what we want to know is, what does greater or lesser UV radiation mean to us? The measurement answer is clear on one level. People who spend a lot of time in the sun end up with roughened and lined skin. This is what gives sun worshippers that folded, old-leather look. While it may frighten some to be told that, if they don't cover up, they are going to end up looking like elephant hide, skin texture is not the key measure equivalent that the skin doctors focus on. What catches everyone's attention is the dreaded word *cancer.* In one sense, the connection has been quite straightforward. Doctors were first able to link the appearance of the relatively benign squamous or basal cancers to lifetime sun exposure at the beginning of the century. And indeed, in the laboratory, ultraviolet radiation has been shown to produce these cancers in laboratory mice. However, it was only in the late 1960s that scientists were able to clearly show a connection between solar exposure and the much more deadly malignant melanoma. Fair-skinned people living near the equator — Australians being the prime example — had significantly higher rates of melanoma than blond or red-haired people living in less sunny northern climes. Confusingly, the increase did not seem to apply to total sun exposure, as outside workers with deep, all-year tans were not particularly at risk. Thus, in the 1970s, the thesis arose that what caused these cancers were sunburns which occurred in young people who didn't yet have tans, but who regularly tried to cultivate them.

The Canadian group also had to deal with the fact that a *de facto* UV measure was already in common circulation. You know it as SPF, or Sun Protection Factor. Creams, lotions, hats, t-shirts, and swimsuits have had SPF levels written onto them, though it is also probably fair to say that people didn't know exactly where they came from. The SPF was officially born in 1978 when the U.S. Food and Drug Administration issued rules that creams and lotions that cosmetic companies advertised as sunburn protectants had to be scientifically linked to something. That something was a variant of a MED. From a physical chemistry point of view, a sunblock of 15 had filtered out all but 1/15 of the UV-B rays. This meant that such a person would be able to be in the sun fifteen times longer than would ordinarily bring on a burn. It is overwhelmingly clear from our earlier discussions that this is probably not a realistic projection. A multitude of factors can influence the amount of radiation that comes through the sky on any given day and how our skin will react to it. Figuring out how long it would be before you burn is almost one of those Chinese puzzles where, if you change one part, the whole thing shifts.

This was the confusing measurement terrain that Canadian scientists had to pull together. Moreover, they had to find a way of delivering the information without creating a panic. After not inconsiderable argument and disagreement, the scientists and public-relations people eventually came up with not one, but three scalings to address the various contradictions between biology, physics, and politics. One was a generalized UV index that was calibrated on a 0-to-10 scale. The 10 was supposedly the peak amount of UV that a person would experience on a clear midday in the tropics. It was modeled on similar scales in place in New Zealand and Australia. Ozone readings and weather information, clouds, sun angle, season, and altitude went into compiling the UV index. Both a daily forecast and hourly readings on the Weather Network on television were available to people trying to plan their sun exposures. The index was itself then translated into two other scales that tried to bring the number into a common person's relationship with the sun. A reading of over 9 was called "extreme." A score of 7 to 9 was called "high." A rating between 4 and 7 was called "moderate," and anything between 0 and 4 was called "low." This was further translated into a figure that a person familiar with a tube of SPF cream could make some sense of. Over 9 meant the famous fair-haired, light-skinned Nordic person would burn in less than 15 minutes; between 7 and 9 was about 20 minutes; between 4 and 7 was about 30 minutes; and under 4 was more than an hour.

If you think this is pretty straightforward, you are wrong. First, if you are at all measure-conscious, you will notice that a significant detail is missing from the scale. What exactly is the unit of measurement which jacks the scale up from one number to another? This was not mentioned, and in terms of subsequent criticism turns out to be vital. When pressed, the Atmospheric Environment Service (AES) scientist said that the number was 25 milliwatts per square meter reaching the Earth's surface. In MED terms, 58 milliwatts would be equal to one of the fair-skinned Nordic types basking for an hour in the midday sun before burning. An increase of 25 milliwatts per square meter was supposed to include a skin burning factor which existed regardless of skin type. A Scottish doctor had come up with just such a measure in 1987. Thus, if you got a burn in 20 minutes at 10, you would get the same burn at 40 minutes when the scale read 5.

Except. Except. Except. Remember all of the caveats we have previously thrown in about human skin and the effect of UV on it. Not to mention the fact that biologic measurement usually has a variability wherein some significant number of people are more than 20 percent above or below the average. And, as the Scottish doctor admitted himself, the vast majority of the data about the effect of UV on skin looked only at pale-skinned Europeans.

What I am saying is that however good the scientists' intentions the biologic weighting system is not all that precise.

Moreover, the new scale had another discombobulation. The top amount of UV on Earth was supposed to be 10. However, as they measured, AES scientists soon realized that the actual figure was much higher. Indeed, a reading of 15 has already been reported in the Canary Islands. Why not bump the scale up then? The answer: the politics of public communication. If the top of the scale was a 15, then average readings in the more densely populated regions of Canada might never get above 5 or 6. "If you have a higher scale, then people think it's only 5 or 6 and get blasé about the whole thing," says AES scientist Jim Kerr. And so, after much debate, it was decided to set a *de facto* Canadian scale in order to achieve the public-health directive of encouraging people to stay out of the sun. So, while it looks as if it is a scale that is simply telling you what is happening in the atmosphere, it is in fact a measure trying to make you do something — in this case, spend less time in the sun.

Now this scale is effective politics, as well as good public-health policy. But it does an injury to good measurement science, particularly now that there is a move afoot to adapt the UV scale to a truly international usage. The problem is that politicized scales can be twisted in various ways. High-altitude tourist meccas in the southern United States and Mexico were almost certain to record readings above the current top of the Canadian scale. "It is fine if you want to frighten people in Canada, but how are you going to make it international enough for people in other countries to use it?" asks Trent University atmospheric scientist Wayne Evans. "The conclusion of the Canadian scale is that the United States is a dangerous place to live," he says, arguing that it would have been better to stick with MEDs. This dissonance has already manifested itself with a vengeance. In June of 1994, the United States Environmental Protection Agency produced its own UV index. It was not a little, but a lot different from the one which the cautious Canadians had so laboriously toiled to create. Because of differences in how they calibrated their measures, there was an intrinsic variance in how the numbers were calculated. A reading of 10 in Windsor, Ontario, was equal to a 9 in Detroit, Michigan — even though the two cities were smack across the Detroit River from one another.

Even worse for border-dwellers trying to make sense of the scales, the U.S. measures tried to take into consideration how clouds might lower the UV reading. The Canadian measure, on the other hand, except for periods of rain or snow, generally didn't factor this in at all. The result, says U.S. meteorologist Craig Long, is that "with our measures, on a day with broken or scattered clouds, a 10 in Windsor, becomes a 5 or 6 in Detroit."

Yikes, you say, this sounds surreal. I'm not done, I respond. The U.S. scale also interprets its numbers differently for the sun-worshipping public. Minimal is 0 to 2. Low is 3 to 4. Moderate is 5 to 6. High is 7 to 9. Very High is everything above 10. Why would you put in place a scale in which "high" in one town is lower than "extreme" in another burg just across the river? National temperament in part. When Americans were surveyed about what words they wanted attached to the scale they found the Canadian terminology too scary. "They answered that 'extreme' was too extreme for them — too aggressive, too. 'High' and 'very high' were more palatable to the public," says Long.

More palatable, no doubt, because the words were less didactic than the Canadian scale — and with good reason. The countries are characterologically different. "Cautious as a Canadian" might almost be the national motto of people living north of the 48th parallel. The United States, on the other hand, was founded by people whose political theory was "that government is best, which governs least." Therefore, all the behavior-altering presuppositions that silently went into making the Canadian scale were viewed skeptically south of the border. "We like to convey information to people and let them make up their own minds; we don't wipe people's noses," opines Long in the best American libertarian tradition.

People involved fear that the result of an "objective" UV scale with two such subjective manifestations will be doubt and scorn. Or, in the worried words of Yvon Deslauriers, a biophysicist with Health and Welfare Canada, "People won't believe the index. They'll say what is going on here is politicians fiddling around, pulling our cords."

There is also the question of the explicit racism in the public face of the index. Why is it, asks sallow-skinned Stephen Strauss, who hardly ever burns, that fair-skinned Nordic folks are the standard? Why isn't each skin type in Canada given its own standard? The answer is that, with all the different measures that could be used and with people's inability to estimate exactly what type they are, it was deemed too confusing. A lot of numbers flummox people. The 0-to-10 scale was decided on because a survey group in Alberta said the original concept of a 0-to-100 scale reminded them too much of Fahrenheit temperatures. But, if the general population ever understands what is going on, there will certainly be an uproar. And I suspect that in the very much more race-conscious United States the boxes that get drawn around the data are going to have to express a greater ethnic diversity.

What, you and Anna's class may now well ask, does this foray into UV mensuration mean? Simple things, I think. There are those who believe the whole effort was politicized into absurdity. "It was just a PR exercise," remarks a caustic Wayne Evans. On the other hand, the UV index, with all its faults, carried precisely the behavioral punch its creators had hoped for. A few months after its introduction, polls showed that nearly 3 out of 4 Canadians were aware that the pain and power of the Sun were now being conveyed to them in a new fashion. Nearly 9 out of 10 thought the tripartite scale a good and understandable idea. And 6 out of 10 said that the scale had made them change their behavior. They were slapping on more sunscreen, slipping on more protective clothing, and avoiding the midday sun.

My own view is more Olympian. After numerous discussions, I don't think there was any conscious political deception afoot among the scale makers. Canadians — and people everywhere — *did* want to know what this ozone scare meant in a simple way, even if it turned out that conveying the true face of nature in all its subtleties was extraordinarily hard. The larger lesson, and maybe it is the largest lesson in all environmental tangles where biology, physics, and politics are gnarled together, is that measurements of natural systems aren't virginal. Nature abhors easy quantification. What we come up with is what someone, or several someones, have chosen to quantify to the exclusion of other evidence. So if we are going to be environmentally wise, we are going to have to be preternaturally skeptical. All the experts' graven truths must be regarded with a hard and doubting eye. When you hear that the toxic waste dump is safe, acid rain is killing maple trees, electromagnetic fields aren't harmful, species are disappearing at a rate of one a day, don't sigh and believe. All these quantifications are suspect because all of them are built on measures whose presuppositions are hidden or obscure. "Statistics is no substitute for judgment," as U.S. politician Henry Clay once remarked.

So doubt, Anna and classmates, doubt and look beyond where the lines of the boxes have been drawn, and propel yourself into that truer place where inherent contradiction forces us to rely on that most basic of survival instincts: common sense.

Part II

The sizesaurus

Intro-duction

Once upon a time — well, about seven years ago — I was in the midst of writing a story for *The Globe and Mail* about an experimental airplane powered by microwaves. It was only a model plane, but the idea of any sort of flying contraption receiving its energy from a ground-based microwave transmitter was exotic, and therefore news. I had the weight of the plane, about 2 kg, but realized sometime late in the day that, since I wanted my readers to "feel" microwaves, not in the usual "cooks an egg in 30 seconds" mode, but as an engine power source of some magnitude, I should translate the weight into something they had some previous experience lifting.

In what I thought at the time was an inspired flash, I decided that I would use the six-pack of beer bottles as my microwave-plane measure. Almost everyone has experienced at one time or other how heavy this is. Simpler said than done. I started calling beer companies in Toronto, but because it was somewhat late in the day, all their public-relations people had gone home. I then started phoning westward, to earlier time zones. Nobody seemed to have the information at their fingertips; everyone was going to have to check with someone else, who may or may not be in. Finally, after several hours of telephoning, I found someone who told me that 2 kg was a little less than two six-packs (a calculation which later proved to be wrong).

I slipped it into the story more than a little angrily. The amount of effort I had gone through to find something so basic seemed to me incommensurate with what I had come up with. What there should be, I said to myself, is some kind of measure thesaurus to which you can turn to find common equivalents of scientific measures — a kind of sizesaurus. And over the years my yearning for just such a reference book increased as I struggled to explain to my readers how small a micron was, how big a light-year, how much pressure 70 pascals exerted, and how acidic something with a pH of 5.4 really was — not to mention grays and centimorgans and Dobson Units . . .

Damn it, I would sputter, why isn't someone writing that sizesaurus? I muttered and muttered and finally realized something which was most unpalatable to a congenitally lazy person: If I truly wanted to have such a book, I was going to have to write it myself. And so I have, with one large caveat and two large acknowledgments. First, the acknowledgments. I have culled information from upward of a hundred books and magazine articles, not to mention including data from helpful people on that massive computer communications system known as Internet. However, two sources stand out. I doubt that I could have written nearly as complete a sizesaurus without *The Guinness Book of Records.* It has, I think, changed the world by giving a face to a thousand, thousand things which curious minds have at one time or other wondered about. But more than that, it has produced a unique cultural image. We are not a very wise age, it seems to me, but we certainly are an age which ecstatically weighs and measures everything. We are the epoch of Top Ten lists and polls and surveys. We may not know who we are, but we have a pretty good idea of how much we are, and that mania for enumeration transfixes anyone reading *Guinness.* It is the bible of an era where quantification is a god.

Second, like Newton, I feel I stand on the shoulders of at least one giant. The pioneering British book *Comparisons* showed me what could be done, how it might be done, and I suppose, equally important, all the ways in which my efforts should be very different from its extremely graphically oriented pages. It is not a sizesaurus, but it is a worthy addition to any measurement library.

Finally, a note on method. Boy, is this sort of book difficult to do. Scientific information is so splintered that to collate it properly one would almost have to look at everything everywhere. I have compromised by looking hard in books that seemed likely to contain significant amounts of information. Some of these referenced their sources; others — college textbooks, for example — often did not. I wrote at least one author of such a text about one of his numbers, and he humbly replied that, when he wrote the first edition twenty years ago, he hadn't realized he should keep notes on sources. Sadly, he now admitted, he had no idea where the number came from. To overcome the curse of secondary sources, I have shown all or part of the "Sizesaurus" to numerous experts in various fields, but I fear that even their intellectual eyes are imperfect. Everything depends on initial assumptions, and if people aren't telling you what those are, you run the risk of extreme error. I frankly fear this, particularly, because if there is anything I have learned in a life in journalism it is that common knowledge is more than commonly wrong.

As this is a book which is backtracking from the modern trend to standardize measurement in ways that everyone agrees upon, I have decided to approach the question of error and mistake in a contrary fashion. I want the next edition of this book to be better than this one, and to that end I have decided to hold a small contest, and to offer a cash reward. Whoever submits to me the largest list of possible miscalculation or mistake by May 1996 will receive a cash reward of $250 and an acknowledgment in the next edition of this book. Documentation must be included, and I am the sole arbiter of whether an objection passes the truth litmus test.

Finally, I hope what I have done serves two quite contrary purposes. One is to allow anyone who wants to translate the language of science into the language of everyman and everywoman to do so with ease. The other is to amuse and entertain. It seems to me that it is possible to read the "Sizesaurus" as a kind of compendium of weird and unusual ways of looking at a world which is marked with a greater precision than we are comfortable with. If any of it makes you laugh, then all of it seems worth it.

The quick-and-dirty word-saurus dictio-nary, or, where to look when you don't want to look long

Note to the unwary: This is not a physics textbook. Nor is it an SI-approved publication. I have included measures which I have come across and struggled to understand, as opposed to every measure in the world. All Imperial measures are based on American numbers on the assumption that they are mainly still used only in the United States. If figures are slightly off, remember that the purpose of the Sizesaurus is to allow people to make relative comparisons. Cross-references to sections of the macropedia are given at the end of each entry.

Acre: One acre equals 4,840 square yards or .405 hectares, which is the equivalent of 11.3 basketball courts or 3.9 Olympic-sized pools. A U.S. dairy cow consumes about an acre's worth of corn silage in a year. (See Area.)

Ampere (A): One ampere (or amp) is the electric current of 1 **coulomb** of electric charge passing a given point in 1 second. People feel a sensation at 0.5 milliamperes; at 50 milliamperes, usually fatal heart convulsions occur.[1] Each car headlight draws around 3 amps, and a toaster between 8 and 10.[2] Named for French physicist, mathematician, and philosopher André-Marie Ampère (1775–1836). (See Electricity.)

Angstrom (Å): The nonmetric measure of the length of things of atomic size. One angstrom equals 10^{-10} m or 0.0000000001 of a meter (0.1 nanometers). An atom is about 3 angstroms in diameter. One sheet of paper is about 1 million angstroms (0.1 mm) thick. One angstrom is ten times smaller than the smallest particles found in smoke. For a scanning tunneling microscope to find a hole 40 angstroms across would be the equivalent of someone finding a pinhead in a haystack measuring 1.93 km (1.2 mi) per side.[3] Named for Swedish physicist Anders Jonas Angström (1814–1874). (See Distance.)

Astronomical Unit (AU): The mean distance between the Sun and the Earth. One astronomical unit equals 1.496 x 10^8 km (9.3 x 10^7 mi). That makes it roughly 22,428 Nile Rivers long. (See Length.)

Planet	Mean Distance from the Sun[4]		
	AU	Kilometers	Miles
Mercury	0.387	58,000,000	36,000,000
Venus	0.723	108,200,000	67,200,000
Earth	1.000	149,600,000	92,750,000
Mars	1.524	228,000,000	141,300,000
Jupiter	5.203	778,300,000	483,600,000
Saturn	9.54	1,427,000,000	887,000,000
Uranus	19.18	2,869,600,000	1,780,000,000
Neptune	30.07	4,497,000,000	2,794,000,000
Pluto	39.44	5,900,000,000	3,658,000,000

Atmosphere (atm): A unit of pressure equal to 101,325 **pascals** or 14.7 lbs per in^2, based on the pressure that the Earth's atmosphere would exert at sea level. A racer's bicycle tire would exert almost 2 atmospheres, and our veins can stand up to 5 atmospheres without bursting.

Avogadro's Number: The number of particles in a **mole** of anything, calculated as 6.022169 x 10^{23}. Named for Italian physicist Amedeo Avogadro (1776–1856).

Equivalents of Avogadro's number include:

1. A hundred billion times the age of the universe in years.
2. The pennies you could give to the 5 billion people on Earth to make each of them a multitrillionaire.
3. The pennies that would make a disk 400 pennies thick the diameter of the Moon's orbit around the Earth.

Bar: A unit of pressure equal to 100,000 **pascals.** Generally cited as millibars, one of which equals 100 pascals. A millibar of pressure is exerted by 1,000 one-dollar bills.

Barn: A tiny area, 10^{-28} m^2, which is used by particle physicists as a target for aiming their atom-splitting beams. Name derives originally from the suggestion that missing an area this size was, atom splitting–wise, like failing to hit a barn door.

Barrel: Barrels come in all sizes. There are 117.3-liter (31-U.S.-gallon) beer barrels, sometimes turned into 119.23-liter (31½-gallon) liquid barrels in some states. There is a 136.26-liter (36-gallon) cistern barrel, not to mention the 159-liter (42-gallon) oil barrels which keep Saudi Arabians in Cadillacs. And oh, did I forget to mention the fruit and vegetable barrels which contain 105 dry quarts or 3.21 U.S. bushels or 113.12 liters? Except for cranberries, of course. Their barrel is 2.709 bushels or 95.46 liters. A barrel of oil would contain about 40 large saucepans or eight North American toilet flushes worth of the black gold. The all-purpose fruit barrel would contain about 45 kg or about 100 lbs of apples. (There is no standard unit size for the apple.)

Baud: The same as a bit or a rate of one pulse per second. Ergo, my modem, which is 19,200 baud, can move information at that rate per second, or about 550 words a second.

Bel: One bel equals 10 **decibels**.

Blink: Roughly defined as 10^{-5} of a day, or 0.864 seconds. An actual human blink takes about 0.33 seconds. (See Time.)

Bohr Radius: The distance between the proton in a hydrogen atom's nucleus and its electron, which equals 5.29177×10^{-11} m. Ergo, a 6-ft man would measure 34,000,000,000 Bohr radii.[5] Named for Danish physicist Niels Bohr (1885–1962).

British Thermal Unit (Btu): The energy needed to raise the temperature of 1 pound of water from 15.5°C to 16.1°C (60°F to 61°F) at one standard atmosphere. It is equal to 1,054.6 **joules** or about 252 **calories**. By way of comparison, the anaerobic breakdown of a **gram** (0.0353 oz) of dried fecal matter will produce 1.4 Btus of heat energy in the form of burned methane. One Btu equals the heat energy given off by a wooden kitchen match.[6]

Bushel (bu): Bushels, like barrels, come in various confusing dimensions. Some bushels are heaped; others are struck (or leveled). The British and the Americans use different measures. A struck U.S. bushel equals 2,150 cubic inches or 32.239 liters. About four of these would fill up a bathtub, and an average 25-year-old man's lungs holds about ⅕ of a bushel. A heaped U.S. bushel is 27.8 percent bigger. So it is a bit more than three heaped bushels to a bathtub. The British struck measure is equal to 1.032 U.S. bushels. So a British bushel basket would hold about 13 kg (2 stone) of apples.

Byte: One byte equals 8 bits. Each letter of the alphabet, basic number, and punctuation mark can be expressed in a byte. (See Information.)

Calorie (cal): (a) The amount of energy required to raise the temperature of 1 **gram** of water by one **Celsius degree**, from 14.5°C to 15.5°C; (b) 1/1,000 of the Calorie (aka kilocalorie) which nutritionists use in computing the energy in foodstuffs. If you took an outer stalk of raw celery about 20 cm (8 in) long and cut it into 10,000 equal pieces, each one of those would contain 1 Calorie.

Candela (Cd): The measure of luminous intensity of the strength of light from a light-emitting or light-reflecting object. It is defined in metric as the "luminous intensity in a given direction of a source that emits monochromatic radiation of frequency 540×10^{12} hertz and that has a radiant intensity in that direction of 1/630 watt per steridian." Replaces the old candle unit, which attempted to quantify the light intensity of actual candles. Still, one candela is roughly the light given out by a candle. (See Light and Illumination.)

Carat: Precious stones, not just diamonds, are weighed in carats, derived from the Arabic word for the seed of the carob tree, or is it the coral tree? Sources disagree. One carat equals 3.086 **grains**, or, more prosaically, 200 milligrams. The largest diamond ever found, the Cullinan from South Africa, discovered in 1905, had a weight of 3,106 carats Out of this, the 530.2 Star of Africa was cleaved. The largest ruby is the 6,465-carat Eminent Star. The smallest diamond had a diameter of 22 mm (0.009 in) and weighed 0.0001022 carats (0.02044 mg) or roughly ⅒ the weight of a middle-sized ant.

Celsius Degree (°C): Originally, 1/100 of the interval between the freezing and boiling points of pure, air-free water at a standard — sea-level — pressure. Now, the equivalent of one degree **Kelvin** or one Kelvin as they say in SI. The freezing point of water is 0°C. Standard body temperature is roughly 37°C. A cool but pleasant day is 22°C; a hot one is 30°C. The coldest temperature, achieved in Antarctica, was –88.3°C. At –40 degrees, the Celsius and Fahrenheit scales converge. Named for Swedish astronomer Anders Celsius (1701–1744). (See Temperature.)

Centimorgan: A mapping unit equal to the distance at which there is a 1 percent probability that a gene has switched from one chromosomal strand of DNA to the other during egg or sperm production. A centimorgan roughly corresponds to a distance of about a million base pairs, or 340 millimeters of DNA for humans. Named for American geneticist Thomas Hunt Morgan (1866–1945).

Chain: (a) An imperial chain is 20.0116 m (22 yds); (2) An engineer's chain is 30.48 m (100 ft); (3) A nautical chain is 4.572 m (15 ft); (4) A square chain is 404.7 m^2 (484 square yds). Belongs in the category of the confusing and is best avoided. (See Distance.)

Coulomb (C): Formally, the charge carried in 1 second by a current of 1 **ampere**. It's a big number. One coulomb of charge is the total charge on 6.24 x 10^{18} electrons. More informally, 2 one-coulomb point charges, if separated by a meter, exert a force on each other of 9 billion **newtons** — roughly the weight of a cube of water a hundred meters (or about 100 yds — one football field on each side).[7] Named for French physicist Charles Augustin de Coulomb (1736–1806). (See Electricity.)

Curie (Ci): 3.7×10^{10} disintegrations a second. The cobalt-60 used in cancer treatments releases about 1,000 curies, and the radium in a fluorescent watch about a millionth of a curie. Named for Pierre Curie (1859–1906; French) and Marie Curie (1867–1934; Polish) codiscoverers of radium. (See Radiation.)

Darwin: A unit of evolutionary rate of change proposed by J.B.S. Haldane in 1948 to measure evolution in terms of the increase or decrease in size or some other genetically significant traits over a million-year period. He estimated that the growth of the modern horse in relation to its forerunners occurred at a rate of 40 millidarwins. Some modern evolutionists are not sure the measurement makes any sense because they believe that the rate of change between species wasn't gradual but relatively abrupt. Named for English naturalist Charles Robert Darwin (1809–1882).

Decibel (dB): The usual unit for measuring the relative loudness of sound, on a logarithmic intensity scale. Ten decibels equals 1 **bel**, and the bel scale increases by powers of 10; that is, 20 decibels is 10 times louder than 10 decibels. Humans can hear sounds over a range of about 130 decibels. Ordinary speech is 60 decibels higher than the lowest sound a person can hear. The base level is defined as the threshold of hearing, which corresponds to a variation in pressure of 3×10^{-5} **pascals**. The threshold of pain, at 120 decibels, corresponds to a variation in pressure of 30 pascals. Atmospheric pressure is about 10^5 pascals. (See Sound.)

Dobson Units (DU): A measurement of the ozone layer; 100 DU equal 1 millimeter (a dime's thickness)[8] of ozone at sea-level pressures. In the atmosphere, the natural range is from 500 DU in the Arctic in the spring to 250 in the tropics all year round. Named for Gordon Miller Bourne Dobson (1889–1976), a British meteorologist. (See Ozone.)

Donkey Power: A unit of power equal to 250 watts; it is computed that 1 donkey power is ⅓ of a **horsepower**, reflecting the relative strength of the two animals. Actually, donkeys can work at peak efficiency at a rate of about ¼ of a horsepower, buffalo and oxen at ⅗ of a horsepower, and horses in brief spurts up to 3 horsepower.[9]

Dram (U.K.: drachm): A unit of weight equal to ⅛ of an apothecaries' ounce, or 60 grains, or 3.88 g. Your standard 325-milligram aspirin tablet weighs 0.0836 drams, and a 500-mg tablet 0.129 drams.

Dyne: Officially, the force that, when applied to a body with a mass of 1 g, gives an acceleration of 1 cm per second per second. One dyne equals 10^{-5} **newtons**, or 1/2,000 of the force the Earth's gravity exerts on a dime. (See Acceleration.)

Electronvolt (eV): A unit of energy used often to express particle energies. One electronvolt equals 1.602×10^{-19} joules. That is too small to say anything about; however, an electron in a TV tube registers at about 20,000 eV. (See Energy.)

Erg: The amount of energy needed to move 1 g through 1 cm with an acceleration of 1 cm per second per second. It is an impossibly small measure of energy; for example, 10,000,000 ergs equal 1 **joule**. Thus, 1 erg is somewhere between the amount of energy you need to make a flea hop and that required for a nerve cell to fire. An erg can be used to show how truly small something is. The energy of a single **photon** is 4×10^{-19} ergs; 60 or so of these must arrive at the eye for it to perceive a flash.[10] Alternatively, at the point at which a compression fracture occurs in a leg bone, the energy being exerted is 19.25×10^{8} ergs.[11] (See Energy.)

Fahrenheit Degree (°F): One Fahrenheit degree equals 5/9 of a degree **Celsius.** It was originally defined as 1/180 of the interval between the freezing and boiling points of water. Now used by Americans to make other people think that their country is the hot spot of the latter half of the twentieth century. Named for German physicist Gabriel Daniel Fahrenheit (1686–1736). (See Temperature.)

Farad (F): Within the metric system the unit of electrical capacitance. One farad equals a condenser (capacitor) which, when charged with 1 **coulomb**, gives a potential of 1 **volt.** A natural condenser is clouds discharging their electricity to the Earth via lightning. A farad is a really big unit. A pacemaker's capacity is 0.4 microfarads or 0.0000004 farads. (See Electricity.)

Fathom: A nautical term which describes a depth sufficient to drown almost all the world's nonswimmers even if they are standing on their tiptoes. It is equal to 1.8288 m (6 ft), which means that there are 880 fathoms in a mile and about 1,547 in a kilometer. (See Distance.)

Femto: A prefix representing 10^{-15}. Smaller than a pinhead's version of a pinhead. A carbon atom's nucleus is about 4 femtometers across; a lead atom's is about 15.

Foot (ft): Originally the length of a royal foot and comprising 30.48 cm (12 in). By modern measures, this means the king wore roughly a size 12 shoe, and might well have been called at the time Big Foot. (See Distance.)

Footcandle (ftc): The illumination of 1 **lumen** over an area of 1 square foot, or the illumination on a surface all points which are at a distance of 1 ft from a directionally uniform point source of 1 **candela.** One footcandle equals 10.76 **lux.** A corridor is dimly lit at 2 or 3 footcandles. (See Light and Illumination.)

Gallon (gal): The British and the Americans used to have their own separate kinds of gallons before the British sensibly became a metric nation. The U.S. gallon is equal to 231 cubic in, or 8 pints, or 3.785×10^{-3} m^3. Generally can be defined as more of anything than you can sensibly drink at one sitting unless you are a camel. (They can drink 30 gallons [113 l] in 10 minutes.) Conversely, it is probably enough gasoline to get a reasonably fuel-efficient North American car from where it is to the next gas station. (See Volume.)

Gauss (G): A measure of a magnetic field's strength. The Earth's magnetic field measures 0.5 gauss. Named for German mathematician, astronomer, and physicist Karl Friedrich Gauss (1777–1855). (See Magnetism.)

Gill: A British unit of volume that equals ¼ of a pint (0.1183 liters). Translation: Enough vodka to give you a good kick, but not enough beer to even make you want a pretzel.

Grain (gr): One grain equals 64.8 milligrams, and is used in the diamond trade when carats are metricized. There are .08 grains to the **carat.** So you could hear jewelers proudly announcing that they stock the finest 45.3593-grain gold watches.

Gray (Gy): The metric unit of energy which radiation imparts to a given amount of absorbing tissue. One gray equals 100 **rads.** At 3 gray, hair loss occurs, and sight is impaired at 5 gray. (See Radiation.)

Hectare (ha): One hectare is 10,000 square meters, and is considered metric's acre. It is one hundredth of a square km or 2.471 acres. An American football field from goal line to goal line is a bit more than two hectares. Vatican City consists of 269 hectares.

Henry (H): The inductance of a closed circle of conducting material which gives rise to a magnetic flux of 1 **weber** per m^2/ampere of current flowing in the conducting circle. You coil a wire, you put a current through, and you get a magnetic field through the center of the coil. To make an inductance of 1 henry from coiled wire with 12 turns per cm and an area of 4 square cm, the coiled wire would have to be 1.4 km (0.87 mi) long. Named for American physicist Joseph Henry (1797–1878). (See Electricity.)

Hertz (Hz): The unit of measurement for the oscillations or cycles per second of any physical quantity. That is, a hertz can be the frequency of electromagnetic waves or soundwaves or water waves. The Kit Kat Clock in my kitchen has a tail which wags back and forth every two seconds and so has a frequency of 0.5 hertz, or ½ cycle per second. Green light is electromagnetic radiation with a frequency of about 10^{15} Hz. Middle C on the piano has a frequency of 262 Hz.[12] Named for German physicist Henrich Rudolph Hertz (1857–1894). (See Sound.)

Horsepower (hp): The power needed to raise 550 lbs 1 ft in 1 second. A metric horsepower is the power needed to raise 75 kilograms 1 meter in 1 second. It is equal to 745.70 **watts.** If you were feeding a horse to produce such energies you would have to give it something more than five times a human's 3,000-**Calorie** daily food intake. (See Energy.)

Joule (J): Officially, the amount of energy needed to move a mass of 1 kilogram through 1 meter with an acceleration of 1 meter per second per second. To put this in perspective, 1,000 **calories** — the "Calorie" which we see in diet measures — equal 4,185.5 joules. One joule of butter would weigh 3.3 x 10^{-6} g, barely enough to see when spread, much less to taste. Another way of describing it: one joule is roughly equal to the amount of energy required to type "required to." Named for English physicist James Prescott Joule (1818–1889). (See Energy.)

Kelvin (K): A scale of measurement beginning at absolute zero (–273.15°C or –459.67°F). If Hell is cold beyond measure, it is colder than 0° Kelvin. Named for British mathematician and physicist William Thompson, 1st Baron Kelvin (1824–1907). (See Temperature.)

Kilogram (kg): The metric mass size most easily translated into daily human needs. One kilogram equals 2.20 lbs, or 1,000 g, or about 16 eggs or 1.5 full bottles of beer. (See Mass.)

Kilowatt-Hour (kWh): A unit of energy denoting 1,000 **watts** of power used or created over the time of one hour. One kilowatt-hour equals 3.6×10^6 joules. In North America, a clock consumes about 17 kilowatts in a year, and a color TV about 502. (See Electricity.)

Knot: One knot is the velocity of traveling 1.852 km (1 **nautical mi**) in one hour, over water usually. (See Distance.)

League: A unit of length, equal to 3,038 **fathoms**, 18,228 ft, 5.556 km, and 3.45 mi. Therefore, if you are out of your league, you are at least 3½ miles from home. (See Distance.)

Light-Year: A measure of very long astronomical distances; it is equal to the distance traveled during one year's time at the speed of light. A light-year is the equivalent of traveling 9,460,528,405,000 km in a year (5,878,499,834,000 mi). The nearest star outside of our solar system is 4.225 light-years away, or 4.00×10^{13} km (2.48×10^{13} mi). One light-year equals the total distance humans were estimated to have driven some kind of vehicle in 1984. The same measure applies for 5 billion people walking 5.2 km a day.[13] (See Distance.)

Link: A unit of length, surveyors link equal to 20.12 cm (7.92 in).

Liter (l or L): The SI unit for volume, equal to 0.001 cubic m, 1,000 cubic cm, 61.02 cubic in, 0.264 U.S. gallons, or 1.057 quarts. A thermos bottle holds about 1 liter. A standard flush toilet flushes about 20 liters of water. (See Volume.)

Lumen (lm): The amount of light given out through a solid angle by a source of 1 **candela** radiating in all directions; that is, 1 lumen is roughly a slice of air being lit up by the light from 1 candle. The lower limit for useful color vision 2 x 10^{-2} millilumens, or 0.00002 lumens. (See Light and Illumination.)

Lux: The intensity of light falling on a surface which is situated, at all points, exactly 1 meter from a point source of 1 **candela.** One lux equals 1 **lumen** per square meter. The daily sunlight reaching the Earth varies from about 60,000 lux in midwinter to 114,000 in summer. (See Light and Illumination.)

Magnitude: A scale for measuring the brightness of stars; also called apparent magnitude. It was first developed by Hipparchus, a Greek astronomer, long before the use of telescopes or modern instruments. He divided the stars into six classes of brightness, with 1 denoting the brightest class and 6 denoting the dimmest class. Nineteenth-century astronomers quantified and precisely defined this scale, and it is still in use today. The range of brightness now extends from the brightest star, Sirius, whose magnitude is –1.5, to the dimmest stars now measurable with magnitudes around 23. The dimmest stars you can see with the naked eye are about magnitude 6.

Star	Magnitude
Sirius	–1.5
Vega	0.0
Regulus	1.4
Betelgeus	0.8
Alpha Centauri (2nd nearest star)	–0.3
Dimmest star visible to the naked eye	6.0
Dimmest star detectable with large telescopes	23
Sun (the nearest star)	–26.74

Meter (m): One meter is now defined as the distance light travels in 1/299,792,458 of a second. One meter equals 39.37 in; it is roughly equal to a large NBA basketball player's stride. It is just a little less than the greatest height from which you can expect your package to be dropped if United Postal Service is handling it in the United States.[14] (See Distance.)

Micron (μ): A micron (literally, a micro-meter) equals one-millionth of a meter. A single fiber in a shag carpet is about 4 to 5 microns across.[15] Tobacco smoke particles are between 0.01 and 1 micron.[16] A hair's width is between 25 and 100 microns. The smallest beach sand is 90 microns, and the largest is 2,000 microns (2 mm) in size. (See Distance.)

Mil (m): When referring to an Imperial measure of length, one mil equals $\frac{1}{1,000}$ of an inch, which is often informally referred to as "one thou." One mil in metric is 2.54×10^{-5} m. An aluminum soft-drink can is about 5 mils thick.

Mile (nautical): A unit of length, over water usually, equal to 1,012 **fathoms**, 6,076 ft, or 1.852 km. Maybe it's longer than a **statute mile** because, generally, you can't go as fast on water as on land, and so a watery mile seems longer than a dry one. (Yes, I know this isn't the real reason.)

Mile (statute): One statute mile equals 5,280 ft, 1,760 yds, or 1.609 km. A medium-long golf course is about 6 km (4 mi) long, and, if you walk briskly, you should able to walk a mile in 10 or 15 minutes on level ground, without stoplights, and without having to pick up after a dog. (See Distance.)

Millimeter (m): A length which is $\frac{1}{1,000}$ of a meter. It is about the thickness of a nickel. (See Distance.)

Mole (mol): Literally, a gram-mole. A mole is the mass in grams of sample that contains the same number of elementary particles (atoms, ions, molecules) as 12 g of carbon. Its numerical equivalent is the atomic or molecular weight of that sample. A mole of water

(H_2O) has an atomic mass of 1 + 1 + 16 = 18, so a mole of water contains 18 g of material, or a little more than a tablespoon. A mole of anything contains exactly the same number of molecules as a mole of anything else. That number is 6.022136×10^{23} — **Avogadro's Number.**

Nano (n): Metric multiplier equal to 10^{-9}. The lowest temperature ever obtained on Earth was 2 nanokelvin above absolute zero, in a low-temperature physics laboratory in Helsinki in 1989.

Newton (N): Formally, one newton is the force required to give a mass of 1 kilogram an acceleration of 1 meter per second per second. Informally, it would feel like the force a medium-sized apple would exert on the palm of your hand. Named for English mathematician Isaac Newton (1642–1727). (See Force.)

Octave: The interval between a fundamental tone and the eighth note above it. Also, all the tones within the interval. One octave note is exactly double the frequency of same note in an octave below it, and half that of the one above it. For example, the note A = 440 vibrations (Hz) a second and A above A = 880 vibrations Hz per second. The range can be seen and heard on a piano where eight successive white keys on the piano equals an octave.[17]

Ohm (Ω): The unit of resistance . . . to just about anything which can be pushed around. There is electric resistance (the resistance to the flow of electric current), acoustic resistance (the resistance to the setting up of soundwaves), mechanical resistance (the resistance to the forcing of a motion), and thermal resistance (the resistance to the flow of thermal energy). The definitive resistance is electric resistance. When ohms are mentioned, this is what is usually meant, unless otherwise specified. Officially, one ohm is defined as the resistance of a circuit in which a potential difference of 1 **volt** produces a current of 1 **ampere.** A newly charged car battery would have an internal resistance of 0.1 to 0.01 ohms. A worn-out battery might have one of between 10 and 100 ohms.[18] Named for German physicist Georg Simon Ohm (1787–1854). (See Electricity.)

Ounce (oz): The ounce is a measure of weight within the Imperial system of measures, and it is also a measure of fluid capacity. One ounce equals 0.0283495 kg ($\frac{1}{16}$ lb). One kilogram equals 35.274 ounces. One fluid ounce equals $\frac{1}{128}$ gallons, $\frac{231}{128}$ cubic inch, $\frac{1}{16}$ liquid pint, 2.957×10^{-5} m^3. Roughly equal to two tablespoons — a bit less if you have a serving made in England. (See Mass.)

Pascal (P): The metric unit of pressure. One pascal equals the pressure exerted by 1 **newton** on 1 square **meter**. The loudest sound people can hear exerts about 30 pascals of pressure. The barometric pressures which weather people report generally weigh in at around 100,000 pascals, or 100 kP. Named for French mathematician and physicist Blaise Pascal (1623–1662). (See Pressure.)

Peck (pk): A unit of volume, equal to 8.80976754272 liters, or about 2 U.S. gallons. I included it here because it is so old-fashioned that it can't help but be charming, especially since there is an English peck which is slightly larger.

Pennyweight (pwt): A unit of mass or weight equal to 1.555 g (0.05 oz). That is equal to one and half paper dollars.

pH: A measure of the acidity or baseness of a substance ranging from 0 to 14, with 0 being the most acidic and 14 most basic. Lemon juice is 2.4, blood is 7, and lye about 13 on the scale. (See pH.)

Photon: The smallest divisible bundle of light, or one "light particle." All light or other electromagnetic radiation is made up of some integer number of photons; the more photons, the more intense the light. The human eye is most sensitive to yellowish-green light, of frequency 5.4×10^{14} **hertz** or photon energy of 2.233 electronvolts. The eye can detect as few as a half-dozen of these little green photons.[19] (See Luminescence.)

Planck: One planck is the action of energy of 1 **joule** for the duration of 1 second. Formally, 1 planck equals 1 J-s. Notice that this is the same unit as Planck's Constant h = 6.6249×10^{-34}, which is a measure of all things extremely small and quantum. For example, one photon of light has an energy 6.6249×10^{-34} times the frequency of the light. Named for German physicist Max Karl Ernst Ludwig Planck (1914–1947).

Point: Used in typography as a measure of the scale of printed letters, one point equals 0.351 mm (0.013837 in).

Pound (lb): The common Imperial weight measure, equal to 16 oz or 454 g. A pound is equal to the amount of weight you can lose in a week if you walk the dog fast and don't ask for seconds on your meals. (See Mass.)

Poundal: Imperial measure's **newton**. The force necessary to accelerate a one pound mass one foot per second per second. It is 13,825.5 dynes or 0.38255 newtons. More prosaically, take a medium-sized apple, cut it into thirds, and call the weight of each third in your hand a poundal apple.

Rad (rd): A measure of the energy which the radiation imparts to a given amount of the absorbing tissue. One rad equals 0.01 **joules** per **kilogram**. The average dosage received by the Japanese at Hiroshima and Nagasaki was 39 rads. Skin burns appear at 300 rads. (See Radiation.)

Radians (rad): (1) A radian is the measure of angle of a circle, with one radian being equal to 57.3 degrees. The normal human eye can just barely distinguish two well-illuminated objects with an angular separation of 5×10^{-4} radians, or about 0.03 of a degree.[20] (2) A radian is also used in the metric system to compute angular speeds. In this measure:

1. The angular speed of the second hand of a watch is 0.1047 radians per second, and of the minute hand, 1.745×10^{-3} radians per second.[21]

2. In 1986 it was reported that the horizontal displacement of the Leaning Tower of Pisa was increasing by 1.26 mm per year. This turns into an average angular velocity of 7.3×10^{-13} radians per second.[22]

Reaumur Degree (°R): A temperature scale formerly used in Europe. The scale divided the temperature interval between the freezing point and the boiling point of water into 80 equal intervals of temperature. One Reaumur degree equals $\frac{4}{9}$(°F –32) = $\frac{4}{5}$°C = $\frac{4}{5}$(°K –273.15). Named for French entomologist René Antoine Ferchault de Reaumur (1683–1757). (See Temperature.)

Rem: A measure of the biological effect of radiation. One rem equals 1 **rad** of 200-kiloelectron-volt X-rays. In the metric system the term for biological quantities is the **sievert**, which equals 100 rems. Exposure to a VDT screen for a year is equal to 0.002 rems. The average minimal yearly exposure of people at sea level is 0.220 rems. (See Radiation.)

Roentgen (R): The measure of the amount of radiation reaching a material. For humans, 600 roentgens would be fatal.[23] Named for German physicist Wilhelm Konrad Röentgen (1845–1923). (See Radiation.)

R-Value: An index of the ability of a substance to obstruct the flow of heat. (See Conductivity, Resistivity.)

Scruple: An apothecaries' unit of mass which equals 20 **grains**, or about 1.3 g. A dollar bill would weigh a little less than one scruple.

Second (s): (1) As a unit of plane angle, one arc second equals 1/(60 x 60) degrees = 0.0002777777 degrees = 1/60 minutes = 0.01666666 = 4.848×10^{-6} radians. A telescope that can see to 1 arc second can see a penny 200 km (125 mi). One that can see $\frac{1}{3}$ arc second can see a penny at 600 km (370 mi).[24] By way of comparison, a peregrine falcon can see a pigeon 8 km (5 mi) away.[25]
(2) As a unit of time, one second equals $\frac{1}{60}$ minute = 1/(60 x 60) hour = 1/(60 x 60 x 24) days.

Sievert (Sv): The SI unit of radiation exposure, equal to 100 **rems**; therefore, the average minimum yearly exposure to people at sea level is about .0022 sieverts. (See Radiation.)

Stere: A unit of volume, used for measuring wood, equal to 1 cubic meter. It was one of the original metric units. A stere would be about the size of freestanding safes which the bad guys are always blowing up in American cowboy movies.

Stone: A unit of mass equal to 14 lbs, or 6.35029 kg (U.K.). People in England still talk about how many stone a particularly big person weighs, but do not describe weight loss in terms of pebbles shed. Funny people, the English, mensuration-wise.

Tesla (T): (See Weber, Gauss.) A unit of magnetism, and consequently understood perfectly only by magnets. One tesla equals 1 **newton** per **ampere**-meter. A refrigerator-door magnet exerts a magnetic force of 0.00002 teslas, the largest man-made magnet 5 teslas. Named for American inventor Nikola Tesla (1856–1943). (See also Magnetism.)

Ton: In Imperial units, one ton equals 2,000 lbs. The metric ton, or tonne, is a different amount of mass, and equals 1,000 kg. The two units sometimes get mixed up because they are pretty close in size. The metric ton is about 200 lbs larger than the Imperial ton. When referring to explosives, it is a measure of energy equal to 5×10^9 **joules**, approximately. A nuclear weapon of power 1 ton has an energy equivalent of 1 ton of TNT. (See Mass.)

Ton of refrigeration: One ton of refrigeration is the rate of extraction of heat when 2,000 lbs of ice of specific latent heat 144 **Btu** per lb is produced in 24 hours from water at the same temperature. In Imperial measure, it is a unit of refrigerating capacity or heat flow.

Torr: A unit of pressure that equals $\frac{1}{760}$ of an **atmosphere**, or about 133.2 **pascals**. Most often used today by doctors. Normal blood pressure is about 120 torr. (See Pressure.)

Volt (V): Formally, it is the electromagnetic force which, when steadily applied to a conductor, is required to make a current of 1 **ampere** flow through a resistance of 1 **ohm.** By analogy, it is sort of like the force required to stuff a continuous tight hot-dog casing with the weird meat they put in hot-dogs. As reference, a car battery produces 12 volts, a flashlight battery 1.5 volts.[26] Named for Italian physicist Count Alessandro Volta (1745–1827). (See Electricity.)

Watt (W): A measure of power, or rate at which work is done, or energy is used up. It is the power provided when 1 **joule** is used for 1 second. A pigeon's metabolic rate is about 1 watt; this means that simply to stay alive, it requires 86,400 joules, or about 20 **Calories,** each day. Named for Scottish inventor James Watt (1736–1819). (See Electricity.)

Weber (Wb): A unit which measures the flux of a magnetic field through a given area. One weber per square meter equals 1 **tesla,** 1 **joule** per **ampere,** or 1 **volt** per **second.** As reference, a refrigerator magnet produces 0.2 webers per square meter; the Earth's magnetic field, 0.0005. Named for German physicist Wilhelm Eduard Weber (1804–1891). (See Electricity.)

A measuring macropedia

A Note to the Reader

Two numerical cautions for those using the "Sizesaurus":

1. Some of the units are expressed in scientific notation, which is both precise and confusing at the same time. Let us consider 10^3. It is another way of saying 1,000, or more precisely 10 x 10 x 10. The confusion arises because a number like 4 x 10^3 is thus equal to 4,000, where, by contrast, 10^4 would turn into 10,000. The minus side of the coin can be just as confusing: 10^{-3} equals .001, or 1 ÷ 10 ÷ 10 ÷ 10. People used to dealing with these exponential numbers seem to make the translations effortlessly, whereas the rest of us either stumble or turn our heads and hope the exponent will turn into something we are more familiar with. Something like 1,000 or 0.001.

2. As noted earlier, Einstein suggested that everything should be as simple as possible but not simpler. Well, nothing speaks to that more than the translation from Imperial to metric measures. In areas where there is bound to be a certain amount of latitude in the measurements — the section on acceleration that follows is an example — I have rounded off some of the Imperial measures. The idea was to expel the notion that the Imperial measure is more accurate; in fact, all the accuracy comes in the measure translation. This is a book infused with the idea that the words *more or less* underlie more or less everything.

Acceleration Deceleration (See Gravity and Force)

Acceleration and deceleration are the increase and decrease, respectively, in velocity occurring from moment to moment. If you are sitting in a car and traveling at a steady speed, there is no acceleration. If you hit the gas pedal, your speed increases over a certain time period. Thus acceleration is recorded in terms of meters (feet) per second per second.

Common Accelerators

Rifle bullet (inside barrel)
■ 3,000 meters per second per second
■ 9,840 feet per second per second

Rat flea jump
■ 2,914 meters per second per second
■ 9,560 feet per second per second

Click beetle jump
■ 400 meters per second per second
■ 1,310 feet per second per second

Squid tentacle extension
■ 330 meters per second per second
■ 1,080 feet per second per second

Locust's (first stage) jump
■ 240 meters per second per second
■ 790 feet per second per second

Lesser galago jump
■ 140 meters per second per second
■ 460 feet per second per second

Spider's jump
■ 51 meters per second per second
■ 165 feet per second per second

Trout's sudden start
■ 40 meters per second per second
■ 130 feet per second per second

Lizard dashing
■ 30 meters per second per second
■ 100 feet per second per second

Antelope jumping
■ 16 meters per second per second
■ 52 feet per second per second

Man's record standing jump
■ 15 meters per second per second
■ 49 feet per second per second

We're All G-Men and G-Women

Subjectively, acceleration is often measured by the effect it has on bodies. This effect is the famous "g" which produces the distortion on the faces of astronauts and test pilots when their planes suddenly increase speed. Scientifically, 1 g is equal to the normal acceleration rate which the Earth's gravity produces on any falling object. That rate is 9.81 m/s^2 (32.17 ft per second per second). Thus, if you want to know something's g, you divide its rate of acceleration by 9.81 m/s^2 (32.15 ft/s^2), a formula which also allows you to read how many gs the creatures are experiencing in the previous table. Ergo:

G, It Hurts

What we are most often concerned about is the effect that accelerations and decelerations have on humans in cars, rocket ships, roller coasters, and the like. Actually the physiological effects differ, depending on the direction of the acceleration and what we are trying to do while the force is operating on us.

Effects of increasing gs in the downward direction (as if you were in a rocket and taking off):

1 g: No noticeable effect in erect or seated position.

2 g: Increase in weight, pressure on buttocks, drooping face and soft tissues.

2.5 g: Difficult to raise oneself.

3–4 gs: Impossible to raise oneself; difficult to raise arms and legs; movements at right angles impossible; progressive dimming of vision after 3 or 5 seconds; progressive tunneling of vision.

4.5–6 gs: Blackout after 5 seconds; bizarre dreams; convulsion in half the people tested; cramps; tingling; difficulty breathing; loss of orientation in time and space for up to 15 seconds postacceleration.

Effects of increasing gs in the upward direction (for example, heading straight downward during a rocket re-entry):

1 g: Unpleasant but tolerable.
2–3 gs: Facial congestion; throbbing headache; blurring or reddening of vision after 5 seconds; eyes puffy; skin hemorrhages.
5 gs: Pass out in 5 seconds.

Effects of increasing gs during forward and backward acceleration
2–3 gs: Increased weight and abdominal pressure; 2 gs tolerable for 24 hours.
3–6 gs: Progressive tightening in chest; loss of peripheral vision; chest pain; difficulty in breathing and talking.
6–9 gs: Increased breathing; chest pain; vision difficulties. At 8 gs forward acceleration, body, legs, and arms cannot be lifted; at 9 gs forward acceleration, head cannot be lifted; and at 9 gs backward acceleration, pain and discomfort occur from pressure of seatbelt or other types of harness.
9–12 gs: Breathing difficulty severe; fatigue; chest pain; loss of peripheral vision; tearing.
15+ gs: Extreme difficulty in breathing and speaking; viselike pains; loss of tactile sensations; recurrent and complete loss of vision.

Movement and Gs: The Harder It Pushes, the Less You Do
With downward pressure
2 gs: Walking, and movement along a ladder difficult.
3 gs: Walking, crawling, and movement along a ladder against acceleration impossible; unaided escape from a vehicle impossible.
5 gs: Difficult to hold feet on rubber pedals.
6–7 gs: Extremely difficult to reach ejection-seat firing mechanism.
8 gs: Arms, legs, and body cannot be lifted.
9 gs: Unsupported head cannot be lifted, although use of counterweight headgear permits motion up to 12 gs.
25 gs: Hand and wrist movement still possible.

EPHEMERA
1. 10 gs — The deceleration force which a redheaded woodpecker's brain is subjected to when it hits the bark of a tree at maximum velocity, i.e., 20.9 km (13 mi) per hour.
2. 110 gs — the temporary deceleration experienced by a diver in Pentecost Island, Vanuatu, when vines stop a free-fall in which the body reaches a speed of 54 km (34 mi) per hour.
3. 179.8 gs — The deceleration experienced by racing driver David Purley when he survived a crash in which his speed decreased from 173 km (108 mi) per hour to 0 in 66 cm (26 in). He suffered 29 fractures, 3 dislocations, and 6 heart stoppages.
4. 400 gs — The acceleration of a click beetle as it jackknifes into the air to escape predators. It is estimated that it endures a peak brain acceleration of 2,300 gs.

Life Is a G		
Approximate Duration and Magnitudes of Brief Accelerations		
Type of acceleration	**Acceleration in multiples of g**	**Duration in seconds**
Elevators		
fast service	0.1–0.2	1–5
comfort zone	0.3	
emergency stop	2.5	
Automobiles		
comfortable stop	0.25	5–8
very unpleasant	0.45	3–5
maximum possible	0.70	3
crash, possible to survive	2–100	0.1
Aircraft		
normal takeoff	0.5	10–20
catapult takeoff	2.5–6	1.5
crash landing, possible survivors	20–100	
seat ejection	10–15	0.25
Humans		
parachute opening	8–33	0.2–0.5
parachute landing	3–4	0.1–0.2
fall into fireman's net	20	0.1

How Gravity on Other Worlds Compares with Earth's			
Planet	**Accelerations**		
	Meters per second per second	**Feet per second per second**	**Gs**
Sun	273.7	898.0	27.9
Mercury	3.6	11.8	0.37
Venus	8.9	29.2	0.91
Earth	9.8	32.15	1
Moon	1.6	5.2	0.17
Mars	3.7	12.1	0.38
Jupiter	26.0	85.3	2.7
Saturn	11.2	36.7	1.14
Uranus	10.5	34.4	1.07
Neptune	13.3	43.6	1.35
Pluto	2.2	7.2	0.23

Area

Area is always a problem, because you often mix what you do know with a measure you don't know. *The New York Times* recently ran an article which described all the failed attempts to translate area into some more meaningful measure. It took to task a pizza-marketing association for suggesting that the 1.8 billion slices of frozen pizza sold each year would cover 1,324,377.1 km² (511,366 square mi). For this to be correct, the *Times* pointed out, each piece would have to cover 744 m² (8,000 square ft). It should also be noted that at least three of the reference books I have consulted don't think it is of interest to their readers to translate square mi into square km (for the curious, the conversion can be made by multiplying the number of square mi by 2.58998811). In any case, the *Times* article went on to describe the various attempts to use the small state of Rhode Island as a *de facto* area metric. Unfortunately, people kept getting its size wrong or missing something in the translation. The equivalent in Canada is the province of Prince Edward Island, which is continually used as a reference, probably again with the same kind of mistakes, particularly if there is any attempt to translate its official square-km size into square mi.

I suggest the following, and caution that every different reference book seems to have its own bases of calculation. These conversions are approximate.

An Area Conversion Scale

Metric	Imperial
1 cm²	0.155 in²
1 m²	10.764 ft²
1 m²	1.196 yd²
1,000 m²	24.71 acres
10,000 m² (hectare)	2.471 acres
1 km²	0.386 mi²

EPHEMERA

1. The surface area of an egg is about 70 cm² (10.9 square in).
2. The head of a flat pin has an area of 1 mm² (0.0016 square in).
3. A flu virus has an area of 10^{-8} mm² (1.6 x 10^{-11} square in).
4. Total skin area of an average-sized man is 1.86 m² (20 square ft) and a woman is 1.58 m² (17 square ft).
5. The total surface area of the 7 m (23 ft) of intestines in humans and all the little bumps and bumps on bumps inside them (villi and millivilli) is 2,000 m² or 0.5 acres.

Large-Scale Geography		
Feature	**Square kilometers**	**Square miles**
Area of Gondwanaland*	2,000,000,000	800,000,000
Total ocean surface	362,033,000	139,781,000
Continental land mass	210,400,000	81,200,000
Pacific Ocean	166,241,000	64,190,000
Atlantic Ocean	82,400,000	31,810,000
Eurasia	53,698,999	20,733,000
Australia	7,618,493	2,942,000
Bay of Bengal	2,172,000	839,000
Hudson Bay	1,232,300	475,800
Red Sea	438,000	169,100
Lake Superior	82,103	31,700
Lake Victoria	69,464	26,828
Lake Michigan	57,757	22,300
Lake Huron	56,600	21,850
Lake Erie	25,667	9,910
Lake Ontario	19,529	7,540
Salt Lake	4,662	1,800

*Giant land mass formed when many of today's continents were mashed together.

Small-Scale Geography			
Site	**Square kilometers**	**Acres**	**Square miles**
The Vatican	0.44	108.7	0.17
The White House	0.073	18.07	0.028
Central Park in New York	3.4	840	1.3125

A Sporting Size		
Playing surface	**Square meters**	**Square yards**
Football field (goal line to goal line)	4,459	5,333
Hockey rink	1,580	1,890
Olympic-sized swimming pool	1,050	1,256
Baseball diamond	753	900
Basketball court	357	427
Tennis court	261	312
Boxing ring	37	44
Pool table	4.6	5.5
Ping-Pong table	4.18	5.0

Human-sized Areas

Parking space for a car
- 12 square meters
- 127 square feet

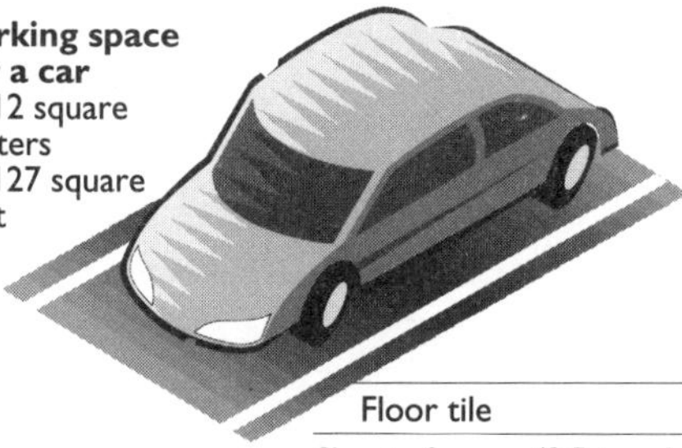

Single bed
- 2 square meters
- 22 square feet

Shower stall
- 1 square meter
- 11 square feet

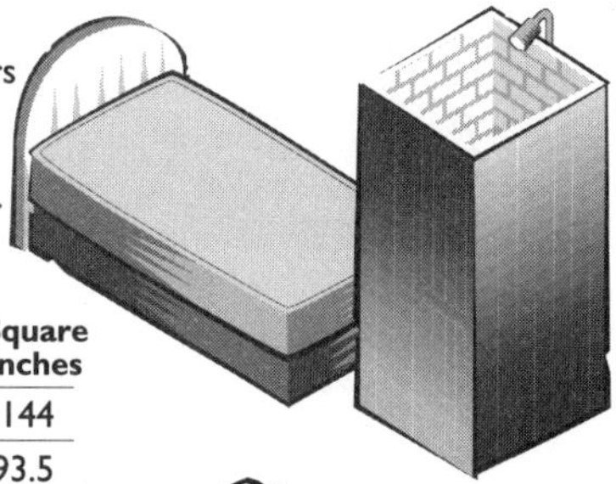

	Square centimeters	Square inches
Floor tile	929	144
Sheet of paper (8.5 in x 11 in)	603	93.5
Standard envelope	190	29.5
U.S. paper currency	104	16.1
Floppy disk (3.5 in)	83	12.9
Playing card	49	7.6
Most credit cards	46	7.2
Standard business cards	45	7.0
Postage stamp	5.5	0.85
U.S or Canadian quarter	4.5	0.70
U.S. or Canadian penny	2.8	0.44

Conductivity (Thermal)

The thermal conductivity of a substance is a measure of the rate at which heat energy can be transferred through its length. But there has to be a temperature difference across the material; that is, it must be hotter at one end and colder at the other. Substances with large thermal conductivities are good heat conductors, and thermal insulators have low values of thermal conductivity. In general, metals are good conductors, and nonmetals are poor conductors.

The metric units of thermal conductivity are watts per meter per degree Celsius. The Imperial measure equivalent is British Thermal Units times inches per hour per ft^2 per degree Fahrenheit.

A Handy Conversion: 1 W/m/°C = 6.933 Btu x in/hr/ft^2/°F		
	Thermal conductivity	
Metals (at 25°C)	**W/m/°C**	**Btu x in/hr/ft^2°/F**
Silver	427	2,960
Copper	397	2,750
Gold	314	2,180
Aluminum	238	1,650
Stainless steel	200	1,390
Brass	103	715
Iron	79.5	552
Lead	34.7	241
Nonmetals (approximate values)		
Ice	1.7	12
Firebrick	1.00	6.93
Concrete	0.8	5.5
Glass	0.8	5.5
Water	0.6	4.2
Rubber	0.2	1.4
Leather in bottom of shoe	0.16	1.1
Asbestos	0.08	0.55
Wood	0.08	0.55

R-Values (or Build a Better-insulated House and the World Will Rip Off Your Siding)

Building and insulating materials are rated in terms of the always-mysterious R-values. In general, all you have to keep in mind is the larger the R-value, the better an insulator the material is. R-value is an Imperial measurement of the insulation's resistance to heat flow. The metric equivalent is the RSI value, which is 0.1761 times smaller than an R-value.

R-Values for Some Common Building Materials	
Material	**R-value (ft^2/°F/h/Btu)**
Fiberglass batting 15 cm (6 in) thick	18.80
Fiberglass batting 8.9 cm (3.5 in) thick	10.90
Brick 10 cm (4 in) thick	4.00
Concrete blocks (filled cores)	1.93
Insulating glass 0.64 cm (0.25 in) space	1.54
Vertical air space 8.9 cm (3.5 in) thick	1.01
Hardwood siding 2.5 cm (1 in) thick	0.91
Wood shingles (lapped)	0.87
Flat glass 0.32 cm (0.125 in) thick	0.89
Drywall 1.3 cm (0.5 in) thick	0.45

Density

Density is a measure of the amount of mass contained in a volume of material. To grasp this concept, use the classic comparison: a ton of feathers versus a ton of bricks. Both weigh the same, but the feathers, which are less dense than the bricks, take up lots more space.

Substance	**Density**	
	Grams per cm³	**Pounds per ft³**
Platinum	21.45	1,338
Gold	19.29	1,204
Lead	11.37	709.5
Silver	10.44	651.5
Granite	2.64	164.8
Concrete	2.16	134.8
Coal	1.35	84.3
Milk	1.03	64.3
Water (at 0°C and 1 atm.)	1.00	62.4
Rubber	0.93	58.0
Petroleum	0.88	54.9
Air (at 17°C)	0.00128	0.08

Bulk Densities of Variable Things, or a More-or-Less Density Scale (The formula at work is the mass of the substance divided by the volume it would occupy.)		
	Kilograms per m³	**Pounds per ft³**
Apples in a barrel	400	25
Beer		
in bottles/cases	449	28
in barrels	529	33
Cork	128–240	8–15
Glass		
bottle	2,723	170
broken	1,522	95
Gold	19,318	1,206
Grain		
barley	625	39
rye	2,643	165
Granite	2,243	140
Ice	913	57
Marble	2,595–2,835	162
Mercury	13,538	845
Mud	1,762–1,922	110–120
Sawdust	208	13
Snow		
fresh	96	6
wet compact	320	20
Tin cans		
cases	481–641	30–40
Tires	176–256	11–16
Water		
fresh	1,001	62.5
salt	1,009–1,201	63–75

Density Odds and Ends		
Animalistic Densities	**Kilograms per m³**	**Pounds per ft³**
Spider silk	1,260	78.7
Red deer antler	1,860	116.1
Cow femur	2,060	128.6
Mollusc shell	2,700	168.6
Human tooth enamel	2,900	181.0
Other Density Measures		
Interstellar space	10^{-18}–10^{-21}	6×10^{-20}–6×10^{-23}
Best laboratory vacuum	10^{-17}	6×10^{-19}
Styrofoam	100	6.2
Platinum	2.14×10^{4}	1,340
Sun at its center	1.6×10^{5}	1×10^{4}
White dwarf stars	10^{10}	6×10^{8}
Uranium nucleus	3×10^{17}	1.9×10^{16}
Neutron star	10^{17}–10^{18}	6×10^{15}–6×10^{16}
Black hole*	10^{19}	6×10^{17}

*A note for purists: Yes, I know the black hole theoretically has an infinite density at its center. But so far black holes only exist in theory, anyway.

Distance (See Size)

Near and far, distance seems different in kind, not just degree. A light-year and a yard are so far apart that, if they were humans, it would be like a Japanese billionaire encountering a starving Somali. Not only would they not speak the same language, but each might doubt the humanity of the other. So here are some tables to guide you in this verbal thicket, but only your imagination can help in making big and small fit on any kind of similar scale.

A Handy Distance Conversion Scale
Imperial to Metric
1 inch = 2.54 centimeters
12 inches = 1 foot = 30.48 centimeters
3 feet = 1 yard = 91.44 centimeters
5.5 yards = 1 rod = 5.029 meters
22 yards = 1 chain = 20.116 meters
63,600 inches = 5,280 feet = 1,760 yards = 320 rods = 80 chains = 1 mile
= 160,934.4 centimeters = 1,609.344 meters
Metric to Imperial
1 nanometer = 0.00000003937 inches
1 angstrom = 0.0000001 millimeters = 0.000000003937 inches
1 micron = 0.001 millimeters = 0.00003937 inches
1 centimeter = 0.3937 inches
100 centimeters = 1 meter = 39.37 inches
1,000 meters = 1 kilometer = 0.621 miles = 49.71 chains = 198.838 rods
= 1,093.613 yards = 3,280.840 feet = 39,370 inches

When Superman Leaps Over Tall Objects, How High Does He Leap?

Object	Height	
	Meters	Feet
Mount Everest	8,848	29,028
CN Tower (Toronto)	553	1,814
Sears Tower	443	1,454 (110 stories)
(with TV towers)	520	1,707
World Trade Center		
North Tower	417	1,368 (110 stories)
South Tower	415	1,362 (110 stories)
Empire State Building	381	1,250 (102 stories)
Eiffel Tower	300.5	985.9
with TV mast	320.6	1,052.3
Washington Memorial	169.1	555
Statue of Liberty		
Sandals to torch	46	151
Pedestal included	93	305
Tallest totem pole	52.7	173
Trajan's Column (Rome)	35	115

A Quick-and-Dirty Relative Scale

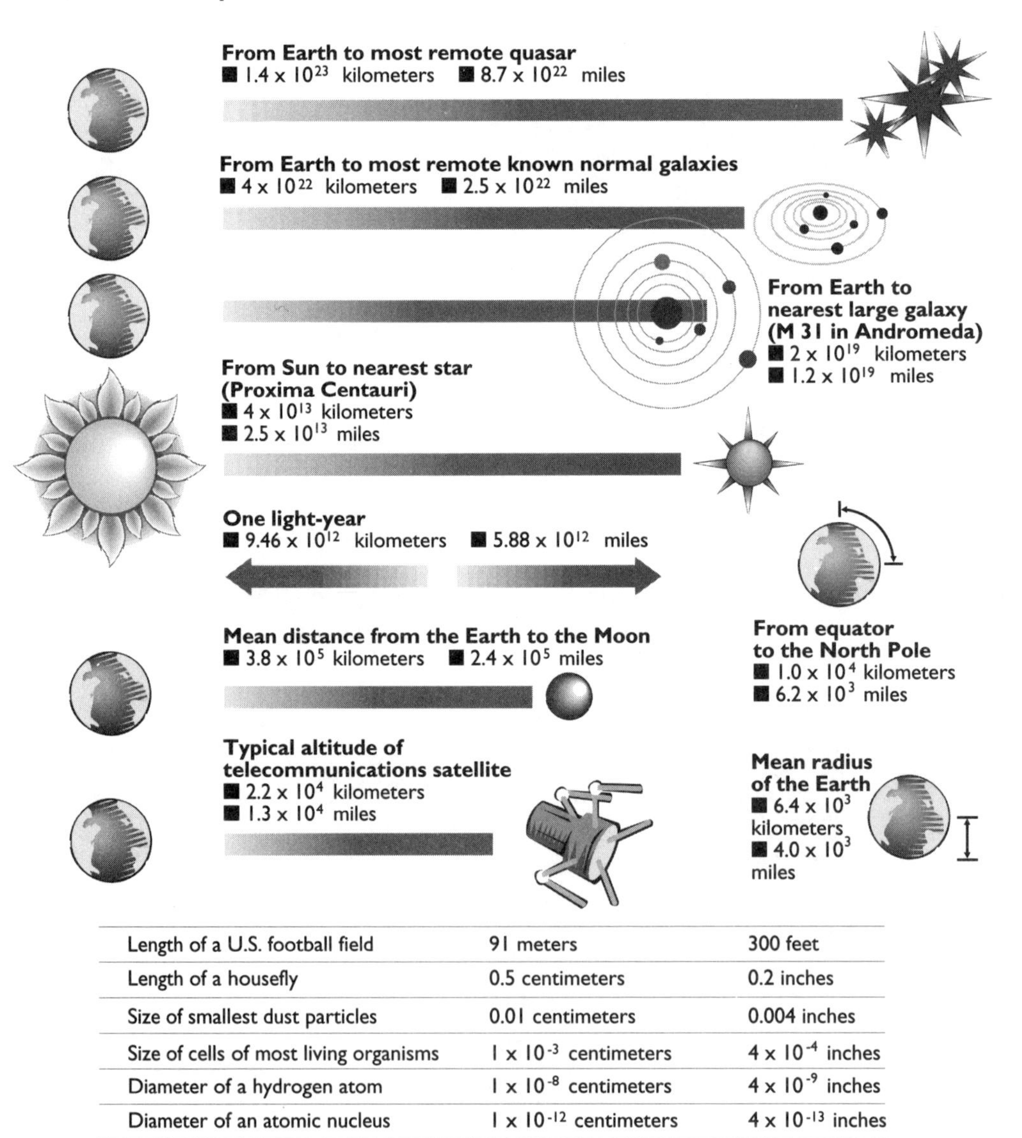

Length of a U.S. football field	91 meters	300 feet
Length of a housefly	0.5 centimeters	0.2 inches
Size of smallest dust particles	0.01 centimeters	0.004 inches
Size of cells of most living organisms	1×10^{-3} centimeters	4×10^{-4} inches
Diameter of a hydrogen atom	1×10^{-8} centimeters	4×10^{-9} inches
Diameter of an atomic nucleus	1×10^{-12} centimeters	4×10^{-13} inches

Height and Depth

How High Is the Sky?

Name	Height
Troposphere	Averages 11 km (7 mi) high. Ranging from 8 km (5 mi) over the poles to 16 km (10 mi) at the equator.
Stratosphere	Ranges from 8 km (5 mi) to 48 km (30 mi).
Mesosphere	Ranges from 48 km (30 mi) to 85 km (55 mi).
Thermosphere	85 km (55 mi) to 700 km (435 mi).
Exosphere	Everything above 700 km (435 mi).
Highest clouds	24,000 m (80,000 ft).
Lowest clouds	1,066 m (3,500 ft).

How Deep Is the Ocean?

Ocean	Average Depth	
	Meters	Feet
Pacific	4,188	13,740
Atlantic	3,735	12,254
Indian	3,872	12,704
Arctic	1,038	3,406
Average	4,000	13,123

EPHEMERA

1. The deepest spot in the ocean: The Marianas Trench near Guam, which is 10,924 m (35,840 ft) deep.

2. Lowest spot on Earth: The Dead Sea, which is 400 m (1,312 ft) below sea level.

3. Highest tide in the world: The Bay of Fundy, at 14.5 m (47.5 ft).

4. In 1987, the number of cigarettes sold in the United States was 4 x 10^{11}. Spread end to end, they would go around the equator 1,000 times.

5. The nearest star to us is 4.22 light-years (4 x 10^{13} km or 2.5 x 10^{13} mi) away.

6. In biblical times, objects, most notably Noah's ark, were measured in cubits. A cubit was originally supposed to be the distance from a man's elbow to the farthest tip of his fingers. The Bible's cubit might be a variety of different distances, depending on whose cubit you are using.

7. Proof positive that Imperial measure is more confusing than metric, or An Inch Divided into Weird Things: An inch, originally the width of a man's thumb, has the following equivalents:
25,400 microns
4,000 silversmith's points
1,000 mils
72.27 points
48 hair's-breadths
23.24635 agates
12 lines
3 barleycorns

Astronomical Comparisons

• If Earth were the size of a period, it would be separated from the Moon by the thickness of one finger (16 mm or 0.63 in), from the Sun by the length of a limousine (6 m or 20 ft), from the nearest star by the length of the Ohio River (1,500 km or 930 mi), from the center of the Milky Way by 200 times the circumference of the Earth (200 x 40,000 km = 8,000,000 km or 5,000,000 mi), and from the Andromeda Galaxy by 4,000 times the circumference of the Earth.

• If Earth were the size of a tennis ball, the crust would be 2.5 times the thickness of the ball's wall; Mount Everest would be barely visible as a bubble; the Nile and the Mississippi–Missouri would be about half the length of a little finger; the troposphere (where weather occurs) would be no thicker than a layer of water on the slightly moistened surface; the Moon would be the size of a marble 2 m (2.2 yds) away and the Sun would be the size of a large 7-m (8-yd) long car parked 700 m (800 yds) away.

• There are about enough atoms in the ink from a letter on this page to provide not only one for each inhabitant of the Earth, but one for every creature if each star in our galaxy had a planet as populous as the Earth.

Cubit	Inches	Meters
Roman	17.48	0.444
Egyptian short	17.72	0.450
Greek	18.23	0.463
Assyrian	19.45	0.469
Sumerian	19.76	0.502
Egyptian royal	20.62	0.524
Talmudist	21.85	0.555
Palestinian	25.24	0.641

A Metaphorical Measure of Distance	
Material	**Distance**
Width of a hair	50 to 75 microns (2 x 10^{-7} to 3 x 10^{-7} in)
Length of a grain of salt	10^{-5} m (0.004 in)
Thickness of a nickel	1 mm (0.25 in)
Length of a thumbnail sketch	1.5 cm (0.6 in)
Diameter of a penny	1.90 cm (0.75 in)
About the amount of yarn needed for each knitted stitch	2.5 cm (1 in)
Length of paper matches	4 cm (1.6 in)
Length of table-tennis table	2.74 m (9 ft)
Distance kangaroos can hop in one leap	7.6 m (25 ft)
Longest frog jump	~10 m (33 ft)
Distance baseball mound stands from home plate	18.44 m (60 ft, 6 in)
Length of the supersaurus, the longest dinosaur	25–30 m (82–98 ft)
Wright brothers' first flight	36.5 m (120 ft)
A stone's throw	50 m (165 ft)
Height of Saturn rocket (used for moon trips)	120 m (363 ft, 8 in)
Length of Noah's ark using Talmudic cubits	166.5 m (546.25 ft)
Longest home run	188.4 m (618 ft)
Length of *QE2* ocean liner	293.5 m (963 ft)
What a single silkworm cocoon contains in silk	3,000–4,000 m (1.9–2.5 mi)
Usual length of lightning channel	5 km (3.1 mi)
Roughly longest length of a lightning channel	100 km (62 mi)
Blood capillaries in every pound of body fat	320 km (200 mi)
Length of Grand Canyon	349 km (217 mi) It is also 6.4–21 km wide (4–13 mi) wide and 1,200–1,650 m (4,000–5,500 ft) deep.
Length of Canada's coastline	90,000 km (56,000 mi)
Length of a blood system	160,000 km (100,000 mi)
Distance of Earth from Sun	149,598,000 km (92,955,900 mi)
Distance Viking Landers traveled to reach Mars	705,000,000 km (440,000,000 mi)
Lengths and widths in the nanometer/micron world	
Length	**Object**
10 nanometers (0.0000004 in)	The diameter of a pipette used to insert sperm into an egg during *in vitro* fertilization
17 nanometers (0.0000007 in)	The smallest virus
4–5 microns (0.00016–0.00020 in)	Width of shag-carpet fiber
10 microns (0.0004 in)	The distance between a cell and a capillary
17 microns (0.0007 in)	The size of a pore in a hen's egg

Einstein's Theory of Relativity

Among the consequences of Einstein's theory of relativity are time dilation and length contraction. Time dilation is the effect wherein, if an onlooker is in a fixed position, a clock moving through space seems to run slower than an identical stationary clock. In length contraction, a moving object seems shorter than an identical stationary object to that same observer. Ergo: The faster the object is moving, the shorter its measurements. Which leads to the following situations:

- At speeds of 100 km (62 mi) per hour, a body contracts by the diameter of one atomic nucleus.
- At a jet plane's speed of 1,000 km (620 mi) per hour, the length of the plane is one atom shorter.
- At an interplanetary rocket's speed of 400,000 km (250,000 mi) per hour, the amount of shrinkage is 0.01 mm (0.0004 in).

Elasticity

Elasticity is a measure of how much pressure you need to apply to stretch or compress things. Like pressure, it is measured in pascals. To give you a sense of that measure, one pascal is roughly the amount of pressure exerted by an ordinary-sized (100 g or ¼ lb) apple turned into apple sauce and spread over an area of one square meter (about the area determined by a very tall man's pace). That is, hardly anything.

Young's modulus of elasticity is the scale of how much pressure you need to apply to stretch or compress things. Young's modulus for a substance is the ratio of applied pressure to the resulting relative change in length in a sample of that substance. Thus, the more fluid a thing, the lower its Young's number, and the stiffer, the higher the number.

Young's Modulus

Material	Mega (millions of) pascals	Pounds per ft^2
Sea anemones' and jellyfishes' jelly	0.001	21
Elastin (nape of neck ligament)	1.0	21,000
Collagen (tendon)	2,000	4×10^7
Silk	4,000	8×10^7
Chitin (outer covering) from a locust	12,000	2.5×10^8
Teeth (dentine)	15,000	3×10^8
Bone	17,000	3.5×10^8
Mollusk shell (nacre)	30,000	6×10^8
Teeth enamel	78,000	1.6×10^9
Steel	200,000	4×10^9

The Breaking Point: How Much Stretching Things Can Stand Before Breaking		
Material	**Megapascals**	**Pounds per in²**
Arterial wall*	2	300
Pine wood across grain (dry)	4	600
Brick, concrete	5	700
Tooth enamel	35	5,100
Tooth dentine	50	7,300
Pine wood with grain (dry)	100	1.45×10^4
Collagen (tendon)	100	1.45×10^4
Bone	200	2.9×10^4
Keratin (human hair)	200	2.9×10^4
Steel (milled)	400	5.80×10^4
Nylon thread	1,000	1.45×10^5
Silk (spider)	2,000	2.9×10^5

*Steven Vogel, in *Vital Circuits* points out that the relative ease with which arterial walls break shows that evolution must not have considered aneurisms a big worry.

Stretchability: A Relative Measure of the Percentage a Material Can Extend Beyond Its Original Length Before Breaking	
Material	**Percentage**
Mollusk shell	0.006
Bone	
Yield or giving strain	0.006
Breaking strain	0.02
Collagen (rat-tail tendon)	0.1
Keratin (wool)	0.5
Elastin (neck ligament)	1.2
Spider silk	0.3–10

Snapping Back: How Easily Something Returns to Its Original Dimensions After Being Stretched	
Material	**Percentage return to original size**
Spider silk	35
Elastin (neck ligament)	76
Collagen (tendon)	93
Resilin (a protein rubber that decelerates and accelerates an insect's wing beats)	97

Toughness: The Amount of Work Per Unit Area It Takes to Fracture Something	
Elastic things bend easily but don't break. Remember, 1 joule in human terms equals the amount of energy a person requires to type 8 letters.	
Material	**Joules/m^{-2}**
Cement	3–40
Wood, with grain	150
Mollusk shell, parallel to surface	150
Tooth enamel	200
Tooth dentine	550
Pig aorta	1,000
Sea anemone jellylike components	1,200
Insect cuticle	1,500
Mollusk shell, crosswise	1,650
Bone (cow leg)	1,700
Steel (high tensile)	10,000
Wood, across grain	12,000
Rabbit skin	20,000
Mild steels	100,000–1,000,000
Fiberglass	1,000,000

Electricity

Electrical measurement is one of those awfully difficult-to-translate areas. Look in first-year physics textbooks and you find yourself awash in equations. If there are real-life examples, they tend to revolve around effects of different wire sizes and imaginary energy uses. That is, much of it is *gedanken* (thought-problem) physics. The good thing for those interested in standardization is that the units are generally all in SI and not Imperial measures.

Electrical Quantities and Their Measure

Volts (V): A measure of electrical force. One volt = 1 joule/coulomb. The potential between the two terminals of a typical flashlight battery is 1.5 volts.

Amperes (A): The rate at which electrical charges flow through a wire. One ampere = 1 coulomb per second. Typical electric currents are in the range of milliamperes to a few amps.

Coulombs (C): The measure of electric charge, a fundamental property of matter (as is mass). Charge can be either positive or negative. One coulomb is equivalent to the total charge of 6.25×10^{18} protons. Another way of visualizing a coulomb: Two 1-coulomb point charges, if separated by a meter, exert a force on each other of 9 billion newtons. Nine billion newtons is roughly the weight of a cube of water a hundred meters (or yards; i.e., one football field) high on each side.

Ohms (Ω): A measure of electrical resistance. That is the property of matter which makes it resist the flow of electric current. It is officially defined as the resistance of a circuit in which a potential difference of one volt produces a current of one ampere.

Watts (W): The measure of the rate at which energy is expended. One watt = 1 joule/sec. A typical household light bulb requires 60 watts of power.

Farads (F): A measure of capacitance or the ratio of the electrical charge to the potential difference between two conductors. One farad = 1 coulomb/volt.

Lightning as an Electrical Measuring Rod

- Electric potential to ground: –10 megavolts from the negative-charge center in the cloud equals +100 megavolts from the positive-charge center in the cloud (which is higher up in the cloud as well) .
- Electric field strength: 10 kilovolts per meter under the cloud.
- Typical voltage drop: 10 kilovolts per meter in the ground or other conductors after lightning impact in the neighborhood.
- Resistance: Several 100 megaohms.
- Peak currents within a lighting stroke: around 10 to 20 kiloamps.

Electric Field Strength of Household Appliances (measured at a distance of 1 foot)

Appliance	Electric field (Volts per meter)	Electric field (Volts per inch)
Electric blanket	250	6.35
Broiler	130	3.30
Phonograph	90	2.3
Refrigerator	60	1.5
Food mixer	50	1.3
Hair dryer	40	1.0
Color TV	30	0.8
Vacuum cleaner	16	0.41
Electric range	4	0.10
Light bulb	2	0.05

EPHEMERA

1. A newly charged car battery would have an internal resistance of 0.1 to 0.01 ohms. A worn-out battery might have one of between 10 and 100 ohms.

2. Wet skin has a resistance of 1,000 ohms; dry skin, 500,000 ohms. The body's internal resistance is only between 100 and 500 ohms.

3. Some other order-of-magnitude charges include orders of magnitude for coulombs:

10^{-6} coulombs: The charge on a typical Van de Graaff generator (the one that makes your hair stand on end at a science center).

10^{-9} coulombs: The charge on a piece of Scotch tape (when pulled off the back of another piece).

4. A toaster uses 10 amps, a TV 3 amps, a car battery 60 amps at 12 volts.

The Amperes of Death: The Effects of Current on a Human Body
Less than 0.01 amp — tingling or imperceptible
0.02 amp — painful, cannot let go
0.03 amp — breathing disturbed
0.07 amp — breathing very difficult
0.1 amp — death due to fibrillation
Above 0.2 amp — no fibrillation, but severe burning and no breathing

Insulators and Conductors, Or Why Wires Aren't Made of Wood

The conductivity of a substance is a measure of how easy it is to set up an electric current in it after applying an electric field. Good electrical conductors have very low resistivity (or a high conductivity), while good electrical insulators have very high resistivity (low conductivity).

Material	Resistivity (Ohm-meters)
Silver	1.59×10^{-8}
Aluminum	2.82×10^{-8}
Iron	10×10^{-8}
Platinum	11×10^{-8}
Lead	22×10^{-8}
Carbon	3.5×10^{-5}
Silicon	640
Glass	10^{10}–10^{14}
Hard rubber	10^{13}
Sulfur	10^{15}
Quartz (fused)	75×10^{16}

Home as an Electrical-Energy Sink Hole

Estimated average annual kilowatt-hours use for a number of common objects in the United States. A kilowatt-hour is a 1,000 watts usage for an hour.

Appliance	Kilowatt-hours
Toothbrush	5
Carving knife	8
VCR	10–70
Hair dryer	14
Blender	15
Clock	17
Shaver	18
Computer	25–400
Garbage disposal	30
Toaster	39
Vacuum cleaner	46
Radio	86
Clothes washer (automatic)	103
Iron	144
Electric blanket	147
Humidifier	163
Dishwasher	165–363
Frying pan	186
Fan (attic)	291
Microwave oven	300
Dehumidifier	377
Television (color)	502
Clothes dryer	993
Range	1,175–1205
Air conditioner	1,389
Refrigerator-freezer	1,591–1,829
Freezer (frost-free)	1,820
Water heater (quick recovery)	4,811

Energy

Energy is everywhere, and according to Einstein's famous $E = mc^2$ equation it is everything. You know, the old matter-is-energy-and-energy-is-matter tautology. The problem is that it is measured a slew of different ways, with perhaps the most confusing fact being that calorie has taken on two meanings. The nutritionists' Calorie is 1,000 of the chemists' calories. Because energy is such a heterogenous arena, this section is full of measures.

Quick and Handy Equivalents
1 joule = 10^7 ergs
1 erg = 10^{-7} joules
1 calorie = 4.185 joules
1,000 calories = 1 kilocalorie or 1 Calorie
1 joule = 6.24×10^{18} electronvolts
1 electronvolt = 1.602×10^{-19} joules
1 kilowatt-hour = 3.6×10^6 joules
1 British Thermal Unit (Btu) = 1,054.5 joules = 252 Calories

Some Kinetic Energies, or the Energy of Moving Things

Event	Joules
Formation of Meteor Crater in Arizona, about 1.2 km (4,000 ft) in diameter and 180 m (600 ft) deep	1.3×10^{18}
Aircraft carrier *Nimitz* 61,400 tonnes (60,400 tons) traveling at 55.5 km/h (30 knots)	4.8×10^9
Satellite orbiting 100 kg (220 lb) at an altitude of 300 km (187 mi)	3×10^6
18-wheeler traveling at 96 km/h (60 mph)	2.2×10^6
NATO SS 109 bullet 4 g (0.14 oz) traveling at 905 m/s (0.51 mi/s)	1.8×10^3
Falling penny 3.2 g (0.1 oz) after 50 m (54 yds)	1.6
Bee, 1 g (0.035 oz) in flight at 2 m/s (6.6 ft/s)	2×10^{-3}
Snail, 5 g (0.17 oz) traveling at 0.05 km/h (0.03 mph)	4.8×10^{-7}
Electron in a TV tube 20 kV	3.2×10^{-15}

Kinetic Energy in Sports

A Fermi Question Energy Scale

A Fermi question is one of the type which Enrico Fermi habitually asked of his students. He wanted them to think about problems involving very large and very small numbers, recognizing that the answers they came up with were likely to be, at best, in the ballpark and, at worst, in the solar system of what the actual number was. (Note: Because so many are so big or so small, they are presented as relative orders of magnitudes; that is, 10^3 will be ten times bigger than 10^2. But the numbers are approximate.)

Energy Flows and Storages	Joules
Big Bang	10^{68}
Radio energy emitted by a galaxy during a lifetime	1^{54}
Energy released in a supernova	1^{44}
Ocean's hydrogen in fusion	10^{34}
Earth's rotational energy	10^{30}
Solar radiation intercepted by Earth per year	5.5^{24}
Global coal resources	2^{23}
Global plant mass	2^{22}
Global annual photosynthesis	2^{21}
Global fossil fuel production	3^{20}
Typical Caribbean hurricane	3.8^{19}
Global lightning annually	3.2^{18}
Largest H-bomb tested in 1961	2.4^{17}
Global zodiacal light annually	6.3^{16}
Latent heat of a thunderstorm	5^{15}
Kinetic energy of a thunderstorm	1^{14}
Hiroshima bomb	8.4^{13}
Coal load in a 100-ton hopper car	2.5^{12}
Good grain corn harvest (8 t/ha)	1.2^{11}
Gasoline for a compact car annually	4^{10}
Barrel of crude oil	6.5^{9}
Basal metabolism of a large horse	1^{8}
Daily adult food intake	1^{7}
Bottle of table wine	2.6^{6}
Lethal dose of X-radiation	1^{6}
Large egg	4^{5}
Vole's daily metabolism	5^{4}
Small chickpea	5^{3}
Baseball pitched at 40 m/s	1.1^{2}
Tennis ball served at 25 m/s	1.5^{1}
Full teacup (300 g) held in hand	2.6
Falling 2–cm hailstone	2^{-1}

Energy Flows and Storages *(continued)*	Joules
Half-dollar falling a meter	1^{-1}
Striking a typewriter key	2^{-2}
Fly on a kitchen table	9^{-3}
Turning this page	10^{-3}
Small bird's 5-second song	5^{-4}
A 2-mm raindrop falling at 6 m/s	7.5^{-5}
The same drop on a blade of grass	4^{-6}
Beat of a fly's wing	10^{-6}
A flea hop	1^{-7}
Discharge of a single neuron	1^{-9}
Energy released by splitting a uranium atom	4^{-11}
0 decibels of sound for one second on eardrums	1^{-15}
Energy to break one bond in DNA	10^{-20}

Humans as Energy-burning Machines

(Remember, these Calories are 1,000 chemical calories.)

Calories Burned Per Hour in Activities

Activity	Man	Woman
Sleeping	65	55
Sitting	90	70
Standing	120	100
Walking	220	180
Walking uphill	440	360
Running	600	420

A Day of Energy Loss

One day's metabolic energy expenditures

8 hours of sleeping = 480 Cal

8 hours of moderate labor = 1,200 Cal

4 hours of reading, writing, watching TV = 320 Cal

1 hour of heavy exercise = 450 Cal

3 hours of dressing, eating = 300 Cal

Energy Densities		
Substance	Energy	
	Joules/kg	Btu/lb
Hydrogen	122,000,000	5.24×10^4
Natural gas	55,000,000	2.36×10^4
Salad oil	37,000,000	1.59×10^4
Walnuts	27,000,000	1.16×10^4
Coal	30,000,000	1.29×10^4
Protein (pure dry)	24,000,000	1.03×10^4
T-bone steak	17,000,000	7,303
Wood, air dry	17,000,000	7,303
Wheat flour	15,000,000	6,444
Lima beans, raw	5,000,000	2,148
Skinless chicken	4,000,000	1,718
Gunpowder	3,000,000	1,289
Pure sugar	3,000,000	1,289
Spinach (raw)	1,000,000	430

EPHEMERA

1. A sedentary human gives off heat at a rate of 100 joules a second, or 100 watts. This means a class of ten is as good at heating a classroom as a kilowatt heater.

2. If sound waves could be converted into electrical power, then 100,000,000,000,000,000 mosquito buzzes could power a reading lamp.

3. At –40°C (–40°F) a person loses about 14.4 Calories per hour by breathing.

4. The total energy consumption needed to make a chick out of

$E = mc^2$ in Gram-sized Chunks

If one could totally convert the amount of energy in 1 g (0.04 oz) of matter, it would be able to:

- Lift 15 million people to the top of Mount Everest.
- Boil 4 liters (a gallon) of water for each of 55 million people.
- Supply a modern city of 15,000 people with electricity for a year.

Talk Energy

• The power output of a man who talks for 3 hours a day through an average lifespan is sufficient to heat up a cup of tea to the temperature it is usually drunk.

• If the speaker's food consumption is 3,000 kilocalories a day, his or her speechifying will be about the energy equivalent of a couple of slices of bread and butter.

• If all the people on Earth (at this juncture, 5.1 billion) spoke at the same time, they would produce 0.46 megawatts of acoustic power.

a fertilized egg is 30 Calories.
5. The average number of extra Calories a pregnant women needs to eat to give birth to a 3-kg (7-lb) baby is 136 Calories per day.
6. The total amount of solar energy it takes to keep the wind blowing is 300×10^{18} joules.
7. A modern electronic digital watch operates on 20 billion times more power than the 10^{-16} volts which are striking Earth from the *Voyager* spacecraft signals 4.5 billion km (2.8 billion mi) from Earth.
8. The amount of power required to keep a person treading water and not drowning is around 13 watts.

Force (See Acceleration, Pressure)

Force, like energy, is so much a part of our lives that it is almost invisible. Starting, stopping, twirling, falling, twisting, and probably just living are all the work of forces. To show you how difficult it is to divorce us from their actions, I once tried to imagine how to construct a golf course on the Moon. The notion was to show people what would happen if two forces — gravity and air resistance — were dramatically altered. However, after a long lunch with a group of physicists, I realized it was almost an impossible problem to simplify. If a golf-playing astronaut swung the same way he or she did on Earth, the backward or forward momentum produced might cause him or her to flip right over. So, perhaps one would have to concoct a whole new type of golf swing. Hitting the ball a long way would be easy, particularly as there is no atmosphere to produce slices and hooks. However, once on the smooth surface — which in the lunar landscape would more likely be called a "brown" than a "green" — putting would be a nightmare. It wasn't clear that the golfer with muscles developed on a planet with a gravitational field six times stronger than the Moon's would be able to hit the ball softly enough. A 1-foot putt might be ten times harder to make than a 100-foot one. Also, because the ball would roll faster (no air resistance) and because so much less gravity was pulling on it, even the truest putts might simply shoot over the hole's mouth. Would you have to make a hole six or ten times bigger than on Earth? Maybe, we decided, the simplest thing was to play some game whose physics didn't change that much. Chess, for instance.

The hardest thing in this book is to know which forces to lump together and which to separate into more easily viewed areas. I have opted for both, so sections on earthquakes, and winds, and acceleration are also really "force" entries. In the metric system, force is measured in newtons. A newton is defined as force acting on 1 kilogram of mass that imparts to it an acceleration of 1 meter per second per second. More metaphorically, one newton is equal to the force which a medium-sized apple would exert on your hand while you stood by the seashore and wondered whether to eat it or throw it. A dyne is equal to the force which accelerates 1 gram 1 centimeter per second per second. Pounds-force are also used in Imperial measure, with 1 pound equal to a little more than four of the previously mentioned newton apples.

An Equivalency Table
1 pound = 4.448 newtons = 4.448×10^5 dynes
1 newton = 10^5 dynes = 0.2248 pounds
1 dyne = 10^{-5} newtons = 2.248×10^{-6} pounds

Forces	Newtons	Pounds
Gravitational pull of the Sun on Earth	3.5×10^{22}	7.9×10^{21}
Gravitational pull of Earth on the Moon	$2. \times 10^{20}$	4.5×10^{19}
Thrust of *Saturn V* rocket engines	3.3×10^{7}	7.4 x 106
Pull of a large tugboat	1×10^{6}	2.2×10^{5}
Thrust of jet engines (Boeing 747)	7.7×10^{5}	1.7×10^{5}
Pull of a large locomotive	5×10^{5}	1.1×10^{5}
Deceleration of automobile during braking	1×10^{4}	2.3×10^{3}
Force between two protons in a nucleus	$\sim 10^{4}$	$\sim 10^{3}$
Accelerating force of an automobile	7×10^{3}	1.6×10^{3}
Gravitational pull of Earth on average person	7.3×10^{2}	1.6×10^{2}
Maximum force upward exerted by a forearm (isometric)	2.7×10^{2}	60
Gravitational pull of Earth on a 5-cent coin	5.1×10^{-2}	1.1×10^{-2}
Force between electron and nucleus of hydrogen atom	8×10^{-8}	1.8×10^{-8}

EPHEMERA

1. The amount of force required to topple a 70-kg (154-lb) person is 45.7 N or 10.3 lbs.
2. Placed at each of the Earth's poles, 2 to 3 g of protons (the weight of a dime) would repel each other with a force of 2.04-teranewtons or 459-billion-lb force.
3. The tension on piano string can be greater than 1,000 N (220 lbs).
4. All of the strings on the grand piano can equal 100,000 N (20 tons) of pressure.
5. The combined tension on the strings of a violin is about 220 N (50 lbs).

Biting, Swinging, Chopping, Smashing

- **220 N (50 lbs) per vertical inch:** The force soft-drink bottles in the United States must withstand without shattering to meet minimum product standards.
- **245–1,245 N (55–280 lbs) of force:** Normal human biting force.
- **~670 N (150 lbs) of force:** The force needed for a karate chop to break a board 1.9 cm (¾ in) thick, by 15.2 cm (6 in) long by 25 cm (10 in) wide.
- **~2800 N (630 lbs) of force:** Centripetal force of a hammer in the hammer throw being swung at 102.4 km (64 mi) per hour.
- **2900 N (650 lbs) of force:** Force needed to break a concrete slab 3.8 cm (1½ in) thick.
- **3600 N (800 lbs) of force:** A karate chop's approximate maximum force.
- **4,339 N (1,062 lbs) of force:** Maximum biting force recorded. Chewing with paraffin wax up to 30 days can increase biting strength by 20 to 25 percent. The teeth of denture wearers exert five to six times less pressure.

Twisted Forces

Breaking torque for human bones	Newtons/m	Pound/foot
Leg		
femur	140	31.5
tibia	100	22.5
fibula	12	~3
Arm		
humerus	60	13.5
radius	20	4.5
ulna	20	4.5

Friction

Friction is the force which keeps everything from continually sliding. If the force being applied is small enough, static friction keeps things from starting to move. Once some force has succeeded in moving an object, its motion is opposed by kinetic friction, the resistance you feel when you rub your hands together. Static friction is always greater than kinetic friction, which is why you shouldn't spin your tires when your car is stuck in the snow. There will be more friction between the tires and the snow if the car is rocking back and forth.

The coefficient of friction is a ratio between the forces opposing the sliding — that is, either static or kinetic friction — and the downward force of gravity. Everything here is relative. The higher the friction numbers the less easily a substance slips along. It is sort of like the conductivity and resistivity of sliding.

EPHEMERA
If Superman, weighing 100 kg (220 lbs), were standing on the ground and trying to stop a 50,000-kg (55-ton) truck traveling at 108 km (67 mi) per hour, the coefficient of friction would mean that, because his ability to stop that truck is limited by his weight, it would take 23 km (14 mi) for him to halt the runaway vehicle.

A Commonsense Table

Surfaces	Coefficients of friction: Static	Coefficients of friction: Kinetic
Steel on steel	0.74	0.57
Aluminum on steel	0.61	0.47
Copper on steel	0.53	0.36
Rubber on concrete	1.0	0.8
Glass on glass	0.94	0.4
Wood on wood	0.25–0.5	0.2
Waxed wood on wet snow	0.14	0.1
Waxed wood on dry snow	0.14	0.04
Metal on metal (lubricated)	0.15	0.06
Ice on ice	0.1	0.03
Teflon on Teflon	0.04	0.04

A way of translating the data in this table might be to say that it is nearly 75 times harder to get steel resting on steel moving than it is to move a similar-sized piece of ice resting on a frozen pond. And once moving, the ice–ice conjunction is something more than 150 times more slippery than steel–steel.

Fuel Economy

Equipment	Operating Distance or Times	Consumption liters / gallons
Voyager satellite	13,000 km (8,100 mi)	1 / 0.26
Toyota LandCruiser	2,049.9 km (1,300.9 mi)	156 / 40
Semitrailer	13.5–14.5 km (8–9 mi)	~4 / 1
M-1 Abrams tank average	1 km (0.6 mi)	47 / 12.4
5-15 jet peak thrust	1 hour	908 / 240
M-1 Abrams tank peak	1 hour	1,113 / 294
F-4 Phantom fighter/bomber	1 hour	6,359 / 1,679
Battleship	1 hour	10,810 / 2,854
B-52 bomber	1 hour	13,671 / 3,609
Nonnuclear aircraft carrier	1 hour	21,300 / 5,620
Carrier battle group	1 day	1,589,700 / 419,700
Armored division (348 tanks)	1 day	2,271,000 / 599,500

Genetics

How do you measure the biology that defines us? Inconsistently, but with great interest.

Long, Thin, and Fully Packed

• In humans, the body's genetic blueprint, the DNA, is roughly a meter or a yard long and probably has somewhere between 2.7 billion and 3.3 billion chemical building blocks (base pairs) in it. Since there are two strands of DNA in each nucleus — one inherited from the mother and one from the father — if all the DNA were stretched out, there would be, in total, somewhere between 2 m (2 yds) of DNA in each cell. A molecule that is a thousand base pairs long would have a length of 0.00034 mm (0.000013 in). All of the DNA is folded and stuffed into a cell's nucleus, which is only 0.005 mm (0.0002 in) in diameter, or 1⁄500 the width of a dime.

• It is often stated that there are between 50,000 and 100,000 genes in DNA — genes are components that send out chemical signals

which actually tell various cells to switch on and off. The appearance of gray hair in aging is one such signal. Unfortunately, nobody really knows how many genes there are in people — guesses have ranged from 30,000 to 300,000. The uncertainty arises because many genes are only about 10,000 base pairs long, while some others have been found that are more than 200,000 base pairs long.
• The vast majority of the DNA – perhaps 90 to 95 percent – is not held as genes but within long sequence of bases which don't seem to activate anything. This means that, if the DNA is a meter or yard long, then only 5–10 cm (2–4 in) of genetic material governs our whole biology.

If DNA Were a Book

At 3 billion base pairs — that is, 6 billion bits of information — there is:
• 21 times more information in a cell than is found in *The Encyclopedia Britannica*, which is thought to have 280 million letters; more than 79 times the information which is found in *The Oxford English Dictionary*, with its 76 million letters; and 500 times more information than is found in the Bible.
• The 46 human chromosomes can vary dramatically in their base-pair lengths. Chromosome 1 has about 300 million base pairs and Chromosome 21 about 50 million.

Genes Tighter-fitting than Jeans

To store the same number of characters as fits into one cell (6 billion) would require 4 floppy disks and a total volume of 40 cm^3 (2.4 cubic in). However, 8 million nuclei can fit into a cubic centimeter (130 million in a cubic inch). And the genetic information packing is denser than that, as DNA makes up only 0.3 percent of the volume of a nucleus. This means that a cubic centimeter of folded DNA contains as much information as 57 billion *Encyclopedia Britannicas.* (A cubic inch is like 930 billion *Britannicas.*)

EPHEMERA

1. Assuming there are 3 billion base pairs in humans, a map of our DNA drawn on the basis of one centimorgan equaling 1,000,000 base pairs would be on a 1-to-3,000 scale. If the beginning and end information for each centimorgan were encapsulated into a 1-in (2.54-cm) long segment, the map would be 76 m (250 ft) long. However, if each base pair were included in a more specific sequencing map, and if they were represented by a color 0.63-cm (¼-in) square, then the map would have to be 18,939.4 km (11,837.12 mi) long.

2. If DNA were the thickness of a clothes-line, it would be 8 km (5 mi) from end to end.

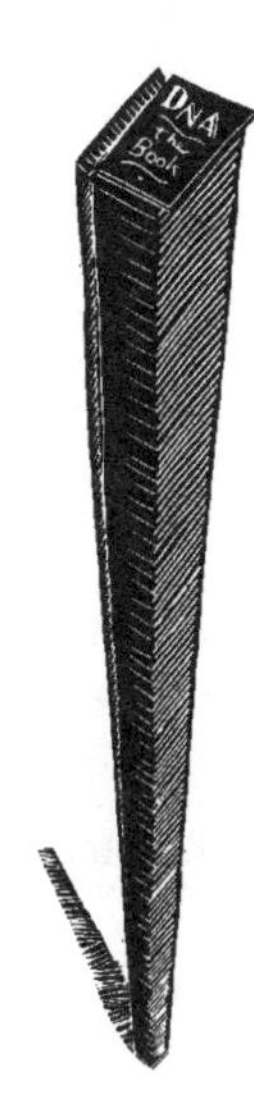

Mapping Terms and Mapping Distance

DNA is mapped either in base pairs or, more usually, in kilobases (1,000 base pairs). However, the centimorgan, also sometimes known as a Mapping Unit, is also used. Formally, a centimorgan is equal to the distance at which there is a 1 percent probability that a gene has switched from one strand of chromosomal DNA to the other during egg or sperm production. This corresponds to a distance of about a million base pairs, or 340 mm (13.4 in) of DNA, although in certain places there may be recombination after as few as 250,000 base pairs.

Gravity (See Acceleration)

We are not just prisoners of gravity — we are its children. Our bodies are shaped to compensate for its pull. If Earth's gravitational constant changed, we would undoubtedly look very different. Make it more, and we would have to become smaller, and maybe denser, to survive. Make it less, and we could conceive of becoming Goliaths.

The Loosening Power

Variations of gravitational pull with altitude	Gravitational pull	
Altitude in km/mi	m/s^2	ft/s^2
0/0	9.83	32.3
5/3.1	9.81	32.1
10/6.25	9.81	32.1
50/31	9.68	31.7
100/62	9.53	31.2
400/250 (space shuttle)	8.70	28.5
35,700/23,562 (satellite)	0.225	0.738
380,000/237,500 (Moon)	0.0027	0.0089

Athleticism as a Prisoner or a Prince of Gravity
If you could jump a meter (a bit more than 3 ft) high on Earth, the same jump would turn into:
3.3 cm (1¼ in) on the Sun
34.5 cm (1ft, 1½ in) on Jupiter
67.7 cm (2 ft, 2¾ in) on Neptune
80.3 cm (2 ft, 7½ in) on Saturn
85.5 cm (2 ft, 9¾ in) on Uranus
101.0 cm (3 ft, 3¾ in) on Venus
240 cm (7 ft, 10½ in) on Mars
250 cm (8 ft, 2½ in) on Mercury
406 cm (13 ft, 4 in) on Pluto
554.2 cm (18 ft, 2¼ in) on the Moon

EPHEMERA
A person who weighed 73.5 kg (162 lb.) at the poles would weigh about 0.5 percent less (368 g or 13 oz less) at the equator. The loss is due to the increased distance from the Earth's center (the Earth has an equatorial "bulge") and the effect of the Earth's rotation: as the Earth spins, an outward (centrifugal) force is felt, slightly counteracting the force of gravity. This effect is greatest at the equator.
There is also a 0.5 percent loss for every 16 km (9.6 mi) ascended above sea level.

Hardness

Hooray. A scale that seems self-explanatory. (*Mohs scale:* Each number represents an increase of a factor of 10 in hardness.)

Diamond	10	
Curundum	9	8–10 will scratch glass
Topaz	8	A knife sharpener = 8–9
Quartz	7	
Orthoclase	6	6–7 will not scratch glass
Apatite	5	A knife blade and teeth = 5–6.
Fluorite	4	A penny = 4
Calcite	3	Bird's egg = 3
Gypsum	2	A fingernail = 2.0–2.5
Talc	1	

Note: This translates into a different relative figure on the Knoop scale. There talc is 1 and diamond is 7,000.

Hardness (Water)

Technically, water hardness equals the total amount of dissolved calcium salts, magnesium salts, iron, and aluminum. Although this should be scientifically measured in parts per million, or milligrams per liter, a variety of other national measures are regularly used, in particular, grains and degrees.

Water hardness is usually most pronounced in human life as it affects washing with soap. There is a simple test you can perform to find out how hard your water is: Take 28 ml (1 oz) of water and add drops of liquid soap to it. The number of drops it takes to produce a soap scum equals the grains-per-gallon on the grains-per-gallon scale. Hardness of water in the United States ranges from 17.1–6,000 mg/l (1 to 350 grains per gallon).

EPHEMERA
1. People cannot generally tolerate drinking water with more than 300 ppm of carbonate, 1,500 ppm of chloride, or 2,000 ppm of sulphate.
2. More than 10,000 ppm of sulphate will cause problems for cows and horses, which generally can tolerate much more brackish water than can humans.

A Handy Translator

1 part per million equals 0.058 grains per U.S. gallons = 0.07 Clark degrees = 0.10 French degrees = 0.056 German degrees. Furthermore, 1 French degree = 1 hydrotimetric degree; 1 Clark degree = 1 grain/Imperial gallon as carbonate; 1 French degree = 1 part/100,000 calcium carbonate; 1 German degree = 1 part/100,000 calcium oxide. My recommendation is stick with parts per million.

This translates into a scale:

Grains per U.S. gallons	Milligrams per liter, or ppm	Type
Less than 1	Less than 17.1	Soft
Between 1 and 3.5	17–60	Slightly hard
3.5–7.0	60–120	Moderately hard
7.–10.5	120–180	Hard
10.5 +	180 and over	Very hard

Information

Computers are machines in which everything is turned into a number. In our era, they quantify quantification for everything else. The computer has demonstrated to the average person in many ways that knowing is counting, and counting is best done by things which aren't human.

In the Beginning There Was the 0 or the 1

From Small —

One bit is equal to on/off, or as computer language reads it, a 0 or 1. It is the smallest unit of information stored in a computer. A byte is 8 bits. A byte is needed to produce a letter or punctuation mark.

— to Large

Amounts of information often slop over into the gigantic. Mega (1,000,000); giga (1,000,000,000); and tera (1,000,000,000,000) prefixes are commonly used, and peta (100,000,000,000,000) lies in the wings.

• A fiber optic wire that carries 1.7×10^9 bits per second corresponds to 25,000 human voices speaking over a wire roughly the size of a human hair.

• A compact disk can store 6 billion bits of information, which is equal to 375,000 pages of text, each holding 2,000 characters.

• It would take 25 terabytes — 25×10^{12} or 25,000,000,000,000 bytes — to store in a computer all the information contained in the Library of Congress's 70 million books, magazines, and newspapers.

• A three-D reconstruction of the brain would require 10^{15} operations, 15 gigabytes of memory, and 1 gigabyte of data access.

• 10^{17} operations, divided into 4 gigabytes of memory and 20 terabytes of data access, equals the amount of information needed to model 40 levels of the global ocean for 100 years.

• To describe all the calculations needed to simulate a standard fusion containment system would require 10^{18} operations, 1 terabyte of memory, and 20 terabytes of data access. As of 1993, 10^{18} operations equal 40 years' operations on a YMP 8/8 computer.

• *The Webster Third International Dictionary* is about 9 cm (3.5 in) thick. It contains 60 characters to a line, 150 lines to a column, and 3 columns to a page, which amounts to 27,000 characters per page. It has 2,600 pages, and therefore about 70 million characters.

Intelligence

"I.Q." stands for intelligence quotient. It is a so-called measurement of intelligence, usually determined by a written test. There is some doubt as to whether a number can characterize a person's intelligence or potential or creative skills, but that doesn't seem to stop us from continually making reference to it. A classic example of how if people desire a measure of something which is hard or impossible to measure, someone will come up with a scale.

Estimates for historical figures:
John Stuart Mill 190, Goethe 185, Voltaire 170, Mozart 150, Thomas Jefferson 145, Napoleon 140, Leonardo da Vinci 135, George Washington 130, Dan Quayle 106.

Light and Illumination

A Handy Reference

One candela = 12.566 lumen
1 footcandle = 10.76 lux

What is hard about light is that it is both specific and diffuse; that is, it begins with some kind of light source and then the light streams off in all directions from it. The diffuse light strikes something else, illuminates that, and — likely as not — is reflected back to begin the process all over again. Beyond this there is question of what we see of all this physics. We turn something neat and exact into mushy biophysics.

Scientists measure light in all its apparitions.

The **candela** measures the total brightness of the source, the total light given out. One candela is roughly equal to the light given off by a candle.

The **lumen** is the amount of light given out through a solid angle by a source of 1 candela radiating in all directions.

Light and Illumination

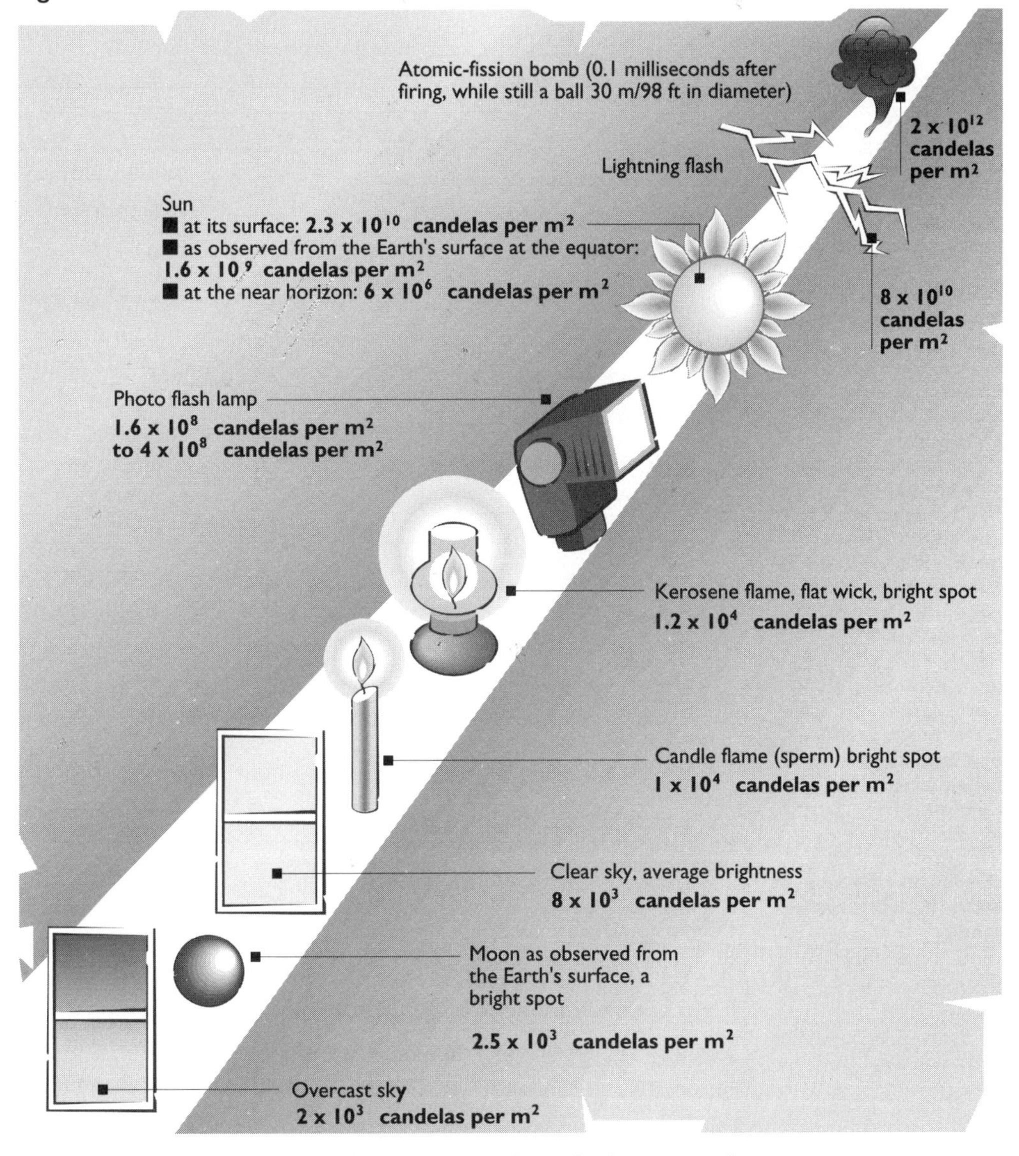

The **lux** is the amount of light falling on a surface which is situated, at all points, exactly 1 meter from a point source of 1 candela. It equals 1 lumen per square meter.

The Imperial measure has traditionally been the footcandle, which is equal to 1 lumen per square foot, or more metaphorically by the amount of light produced by a candle on a surface at a distance of 1 foot.

EPHEMERA
1. In a light fog, a brightness of 9,000,000 candelas is required for a light source to be seen 4.8 km (3 mi) away.
2. A corridor is dimly lit at 20–30 lux. A non-reading room is adequately lit at 200 lux. Reading requires 500 lux, precision drafting 2,000 lux.
3. The average brightness of the Sun is 1,600 megacandelas. The illuminance on the Earth's surface on a clear day can exceed 100 kilolux, or 10,000 footcandles. On a cloudy day, that drops to 10 kilolux, or 1,000 footcandles.
4. The Moon's reflection is 2,500 candelas per square m (16,000 candles/in²). The illumination on the Earth's surface on a clear night may be 0.1 lux, or 0.01 of a footcandle.

How Bright Are Night and Day	
Horizon sky	**Candelas/m²**
Overcast, no Moon	0.00003
Clear, but no Moon	0.0003
Overcast, Moon	0.003
Clear, moonlight	0.03
Twilight	3
Very dark day	30
Overcast day	300
Clear day	3,000
Daylight	
Dull	300–1,000
Typical	1,000–3,000
Bright	3,000–16,000
Ground	
On an overcast day	30–100
On a sunny day	300
Snow, full sunlight	16,000

White as a Rainbow

So-called white sunlight is in fact composed of a variety of wavelengths. They are measured in billionths of a meter — nanometers. The light we see ranges in wavelength between 400 and 700 nanometers, with violet 390–455, blue 455–492, green 492–577, yellow 577–597, orange 597–622, and red 622–770. The eye cannot see the ultraviolet rays, which are between about 180 and 400, and infrared, which are from 770 all the way to 1 millimeter.

Light as Energy

The smallest divisible bundle of light, or one "light particle," is the photon. All light or other electromagnetic radiation is made up of some integer number of photons; the more photons, the more intense the light. The energy of one photon varies, depending on the wavelength, or color, of the light or electromagnetic radiation. For example, one photon of violet light has an energy of 3.1 eV (electronvolts), and one photon of red light has an energy of 1.8 eV. One eV is a unit of energy equal to 1.6×10^{-19} joules. The human eye is most sensitive to yellowish-green light, of frequency 5.4×10^{14} Hertz or photon energy 2.23 eV, and can detect light when as few as a half-dozen of these little green photons are present.

Types of Electromagnetic Radiation and the Corresponding Energies of the Individual Photons

Name of electromagnetic radiation	Energy (joules)	Photon energy (electronvolts)
Radio waves	less than 2×10^{-22}	less than 0.0012
Infrared light	2×10^{-22} to 2.8×10^{-19}	0.0012 to 1.8
Visible light	2.8×10^{-19} to 5×10^{-19}	1.8 to 3.1
Ultraviolet (UV) rays	5×10^{-19} to 2×10^{-17}	3.1 to 120
X-rays	2×10^{-17} to 2×10^{-15}	120 to 12,300
Gamma radiation	greater than 2×10^{-15}	greater than 12,300

Best and Brightest

• Brightest light: A picosecond laser produced a light with an intensity of 5×10^{15} watts.

• The most powerful searchlight had an arc luminance of 46,500 candelas per square cm (or 300,000 candles per square inch). It produced a maximum beam intensity of 2.7 billion candles when reflected in a 3.05-m (10-ft) parabolic mirror built by the General Electric Company in London during World War II.

• The most powerful continuously burning light source, an argon arc light, produces 1,200,000 candelas.

A Relative Measure of Sight		
Distance one can see in different weather as determined by the atmosphere's ability to make signals diminish.		
Viewing conditions	**Kilometers**	**Miles**
Exceptionally clear	50 +	31+
Very clear	50	31
Clear	20	12
Light haze	10	6
Haze	4	2.5
Thin fog	2	1.2
Light fog	1	0.6 (3,200 ft)
Moderate fog	0.5 (500 m)	0.3 (1,600 ft)
Dense fog	0.05 (50 m)	0.03 (160 ft)
Very dense fog	0.03 (30 m)	0.019 (100 ft)
Exceptionally dense fog	0.015 (15 m)	0.009 (50 ft)

Seeing on the Millilumen Scale

- Lower limit for useful color vision is 2×10^{-2} millilumens.
- Upper limit for night vision is 1×10^{-3} millilumens.
- Absolute threshold for dark-adapted human eye is 1×10^{-5} millilumens.

Magnetism

There is an interplay of units which relate to magnetic fields. Like electricity, it is one of those areas of science where it is easier to describe what is happening mathematically than conceptually, because electromagnetic fields are subject to many different laws. Although the tesla is the official SI magnetic unit, the gauss is regularly used and cited.

Handy Equivalents

1 tesla equals 1 newton/per ampere-meter

1 tesla is also equivalent to 1 weber per meter^2

1 gauss equals 0.0001 teslas

Magnetism, Object by Object		
	Gauss	Tesla
Weakest lab field	8×10^{-11}	8×10^{-15}
Brain's magnetic field	2×10^{-8}	2×10^{-12}
Interstellar galactic space	10^{-6}	$\approx 10^{-10}$
Produced by human body	3×10^{-6}	3×10^{-10}
In sunlight (rems)	0.3	0.00003
Milky Way's magnetic field	0.2	0.00002
Earth's magnetic field	0.5	0.00005
Near household wiring	0.1	0.0001
In a sunspot	3,000	≈0.3
Largest man-made magnet	50,000	5.0
Strongest continuous field	355,000	35.5
At surface of a pulsar	≈1,000,000	1,000
At surface of a nucleus	≈10,000,000,000	1,000,000

Power-Frequency Magnetic Fields of Household Appliances	
Appliance	**Range in gauss**
Soldering gun/hair dryer	10–25
Can opener/electric shaver/kitchen range	5–10
Food mixer/TV set	1–5
Clothes dryer/vacuum cleaner/heating pad	0.1–1
Lamp/electric iron/dishwasher	0.01–0.1
Refrigerator	0.001–0.01

Electromagnetic Fields We Use and Abuse

The fields below are grouped according to frequency in megahertz. Remember, a single hertz is one cycle — that is, a complete to-and-fro motion of an electromagnetic wave. Technically, a hertz equals 1 cycle per second. A megahertz equals 1,000,000 cycles per second.

Electromagnetic Field	Megahertz
Microwave relay, short-range military communications	300,000
Commercial satellites, microwave relays, radar air navigation, military communication	30,000
UHF TV, police and taxi radios, microwave ovens, medical diathermy, radar, weather sats.	3,000
FM radio, VHF TV, police and taxi radios, air navigation, military satellites	300
International shortwave, air and marine communication, long-range military communication, HAM radio, CB	30
AM radio, air and marine communication, HAM, SOS signals	3
Air and marine navigation	>0.03
Time signals, military communication	0.03
Electric power, military com, bone stimulation, electric transport systems	0.003
Electric power, batteries, bone stimulation	0

Mass (See Gravity)

Weight and mass, mass and weight — it's always confusing to the nonscientist, and you are almost always certain to misuse the terms. *Mass* is a property of a body itself. It's an intrinsic quality, a measure of its inertia or a measure of its quantity of matter. *Weight* is the gravitational force acting on an object. What the physicists are trying to do is differentiate what we seem to weigh at sea level on the Earth as opposed to what we would seem to weigh if we were on the Moon, or walking on the surface of the Sun. Our mass stays the same, but how that mass manifests itself in heaviness doesn't. To differentiate between the two, different measures are used. A newton or, in Imperial measure, a pound-force or poundal, is used to describe weight. In the SI system, the kilogram, and still in Imperial measure, a pound, is used to describe gravitational mass. The translations are 1 newton = 0.225 pound-force, and 2.25 pounds are just about 1 kilogram. Ergo, a fairly heavy man on Earth might weigh 1,000 newtons, or 225 pounds.

Masses, Smallest to Largest		
Object	Representative masses	
	Kilograms	Pounds
Graviton	10^{-70}	2×10^{-70}
Photon	5.3×10^{-63}	1.2×10^{-62}
Neutrino	3.2×10^{-35}	7.0×10^{-35}
Electron	9×10^{-31}	2.0×10^{-31}
Oxygen atom	3×10^{-26}	6.6×10^{-26}
Insulin molecule	10^{-23}	2×10^{-23}
Penicillin molecule	10^{-18}	2×10^{-18}
Period in this book	6×10^{-9}	1.3×10^{-8}
Giant amoeba	10^{-8}	2×10^{-8}
Ant	10^{-5}	2×10^{-5}
House spider	10^{-4}	2×10^{-4}
Hummingbird	10^{-2}	2×10^{-2}
Dog	10	22
Human	10^{2}	220
Elephant	10^{4}	2×10^{4}
Blue whale	10^{5}	2×10^{5}
Oil tanker	10^{8}	2×10^{8}
Weight of the ocean	1.32×10^{21}	2.9×10^{21}
Moon	7×10^{22}	1.5×10^{23}
Weight of air at sea level	10^{24}	2×10^{24}
Earth	6×10^{24}	1.3×10^{25}
Jupiter	1.90×10^{27}	4.19×10^{27}
Sun	2×10^{30}	4×10^{30}
Our galaxy	2×10^{41}	4×10^{41}
Mass of the observable universe	10^{51}	2×10^{51}

By Analogy

- If the Earth were broken into pieces and transported on freight trains, a 20-ton freight car loading at one car a second would take 1,000,000,000,000,000,000 years to clear away the rubble.
- In terms of their masses, an electron is to a watermelon as a watermelon is to the Sun.
- A North American man weighing 72 kg (164 lbs) could be a measure of other objects: 1,300 of such men equal an average blue whale; 136 equal an average elephant; and 5.9 equal an average horse.

The Common Touch

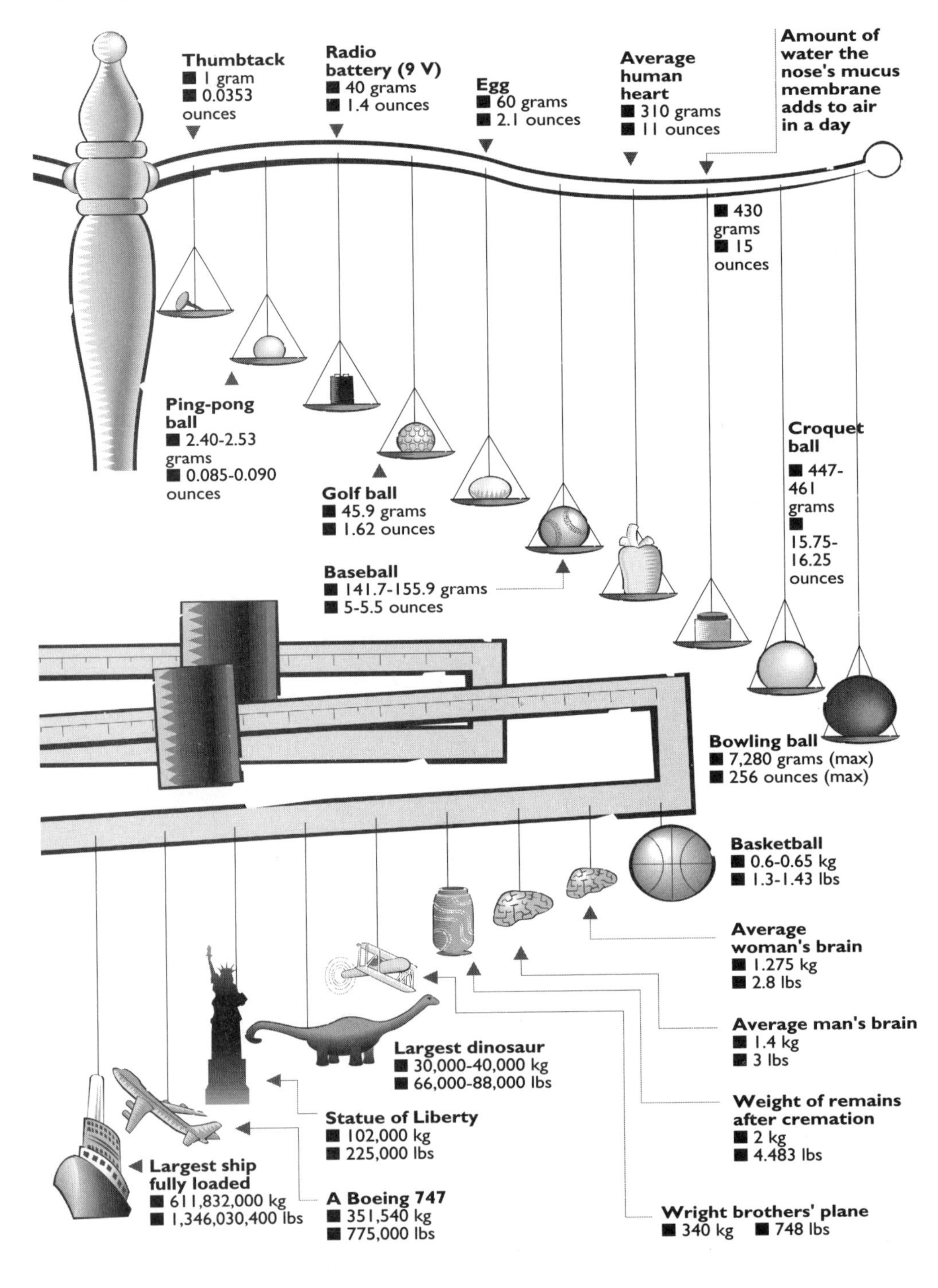

Weight by Money		
Object	**Grams**	**Ounces**
Bill	1	0.0353
Dime and penny	2.5	0.088
Nickel	5	0.176
Quarter	6.25	0.22
Half-dollar	12.5	0.44

Animal Weight Metric		
Animal	**Grams**	**Ounces**
Pygmy shrew	4.53	0.16
Mouse	22.40	0.79
Goliath beetle	85	3
Golden hamster	119	4.2
Guinea pig	680	24
	Kilograms	**Pounds**
Rabbit	3.6	8
Chicken (heaviest)	10	22
Cat	6.3	14
Racoon	9.5	21
Porcupine	27	60
Coyote	34	75
Porpoise	47	103
Alligator and chimp	68	150
Mountain lion	77	170
Wild boar	137	307
Llama	170	375
Moose	360	790
Pilot whale	680	1,500
Walrus	1,450	3,200
Cow (largest)	2,267	5,000

EPHEMERA

1. The total amount of ragweed pollen which is inhaled during a season in the United States is between 100 and 3,000 micrograms (0.0001–0.003 g or 0.000004–0.0001 oz) and it may contain as little as 1–3 micrograms (0.00000004–0.0000001 oz) of active allergenic substance. This translates into a few nanograms (0.000000001 g or 0.00000000004 oz) per sneeze.

2. 4.7×10^{-9} g or 1/600,000,000,000 oz is how much grass pollen weighs.

3. 0.000001 g or 0.00000004 oz is the weight of a snowflake.

4. 0.0049 g or 0.00017 oz is the approximate weight of a flea.

5. 0.001 g or 0.00004 oz is the weight of the brain of a bee.

Heavyweight and Lightweight People		
	Kilograms	**Pounds**
Man		
Most	635	~1,400
Woman		
Most	399	878
Least	5.9	13

Mechanical Heavyweights		
Vehicle	**Tonnes**	**Tons**
Tractor	7.44	8.2
Greyhound Bus	12.15	13.4
18-wheel truck (loaded)	44	48.5
Chieftain tank	52.8	58.2

Numbers (Large) (See also Odds, Earthquakes)

The metric system has a series of prefixes to express big and little numbers that are found everywhere in science.

yotta	10^{24}	deci	10^{-1}
zetta	10^{21}	centi	10^{-2}
exa	10^{18}	milli	10^{-3}
peta	10^{15}	micro	10^{-6}
tera	10^{12}	nano	10^{-9}
giga	10^{9}	pico	10^{-12}
mega	10^{6}	femto	10^{-15}
kilo	10^{3}	atto	10^{-18}
hecto	10^{2}	zepto	10^{-21}
deca	10^{1}	yocto	10^{-24}

Millions (10^6–10^8)

- There are 6,000,000 (6×10^6) smell cells in our nose.
- It would take 500,000,000 gold atoms (5×10^8) to stretch across the 15.7-cm (6.18-in) length of a dollar bill. It would take 210,000,000 (2.1×10^8) atoms to cover its width.

Billions (10^9–10^{11})

• There are 3 billion (3×10^9) possible ways chess players can play the first four moves in a game.

• Five billion (5×10^9) equals the number of matches which could be made out of the largest tree in California — 83.3-m (273-ft) tall General Sherman.

• When we flush the toilet, somewhere between 5 billion and 10 billion (5×10^9 and 10×10^9) water droplets go into the air.

• About 70 billion (7×10^{10}) people have existed on Earth.

Trillions (10^{12}–10^{14})

• The energy content of an extremely loud sound is a trillion (10^{12}) times greater than that of the weakest sound that can be heard.

• There are 2,230,000,000,000 (2.23×10^{12}) grains of sand on the surface of New York's Jones Beach.

• We have 25 trillion (2.5×10^{13}) red blood cells in our body at any one time.

• The largest concentration of animals was 12,500,000,000,000 locusts, weighing 25,000,000 tonnes (24,600,000 tons), which appeared in Nebraska in 1875.

Quadrillions (10^{15}–10^{17})

• There are about 1,000,000,000,000,000 (10^{15}) ants on the Earth.

• It would take 10.5×10^{16} atoms to cover a U.S. dollar bill entirely.

• It is estimated that the oceans contain 5×10^{17} tonnes (4.9×10^{17} tons) of salt. If spread out over the Earth's surface, it would create a layer 152 m (500 ft) thick.

Larger Still

10^{18}	The number of molecules in a snowflake.
10^{19}	There are 17,000,000,000,000,000,000 (1.7×10^{19}) water molecules in a drop of sea water.
10^{21}	In four months, a pair of flies could produce 1,910,000,000,000,000,000,000 (1.91×10^{21}) descendants.

10^{21} The global net annual photosynthesis in joules.

10^{22} There are 60,000,000,000,000,000,000,000 (6×10^{22}) molecules in an average raindrop.

10^{23} See Avogadro's Number.

10^{24} Solar radiation intercepted annually by the Earth in joules. Also the number of bacteria it would take to fill the volume of a large whale. Laid end to end, these bacteria could circle the Earth 7.6 billion times, go to the Sun 1 million times, or go to the nearest star — 3.3 light-years away — and return about 5 times.

10^{27} It would take 10^{27} sugar cubes to equal the volume of the Earth; 3×10^{27} is roughly the number of oxygen atoms in an average-sized North American bedroom.

10^{28} The number of *E. coli* bacteria (found in the human stomach) which could exist after a day if one bacterium was allowed to reproduce in ideal circumstances.

10^{29}–10^{30} The Earth's rotational energy in joules.

10^{35} The estimated number of snowflakes which have fallen on Earth since it was formed.

10^{40} The orbital motion of the Earth in joules.

10^{43} The orbital motion of all planets in the solar system (3.2×10^{43} joules).

10^{44} The number of water molecules in the Mediterranean; the energy released in a supernova in joules.

10^{51} An estimate of the number of years that it would take for the light touch of an Indian holy man, who touches it only every million years, to wear away a legendary cubic stone, 1.6 km^3 high (a mile high), which is a million times harder than diamonds.

10^{67} There are 806 million, billion, billion, billion, billion, billion, billion (8.06×10^{67}) possible arrangements of 52 playing cards.

10^{68} The amount of energy in joules released by the Big Bang.

10^{79}	Or 15,747,724,136,275,002,577,605,653,961,181,555, 468,044,717,914,527,116,709,366,231,425,076,185,631, 031,296 — Sir Arthur Eddington's wrong estimate of the number of protons in the universe.
10^{80}	The number of atoms in the universe.
10^{87}	The number of electrons in the universe.
1 in $10^{41,600}$	The likelihood that a given monkey could accidentally write *Hamlet.*
$10^{75,000}$	A number bigger than number of possible chess games.

EPHEMERA
1. The number of cells in the average leaf is 40,000,000.
2. The average number of quills on a porcupine is 25,000.
3. The average number of thunderstorms in a year is 16,000,000.
4. There are about enough atoms in the ink that makes a letter in this sentence to provide not only one atom for each inhabitant of the Earth, but one for every creature if each star in our galaxy had a planet as populous as the Earth.

Body Size Metric

- Redheads have about 170,000 hairs on their head, blonds 185,000, and brunettes and black-haired people about 200,000.
- People can hear 500,000 different sounds.
- The body has 2 million to 3 million sweat glands.
- There are about 5,000,000 hairs on the average human body.
- There are 7 billion capillaries in the human body.
- There are 25 trillion red blood cells in the body.
- There are 60 trillion cells in the body.

Odds (See also Large Numbers)

Odds are incredibly localized, with chances of death and dismemberment being a function of the kind of society you find yourself in. In 1991, Clark Chapman, an astronomer with the Planetary Science Institute in Tucson, Arizona, achieved instant media notoriety when he produced a chart in which he tried to quantify the chances that an American would be killed by an asteroid/comet collision versus more commonly perceived threats.[1]

It read:

Cause of Death	Chances
Cigarette smoking	1 in 7
Motor vehicle accident	1 in 100
Homicide	1 in 300
Fire	1 in 800
Firearms accident	1 in 2,500
Electrocution	1 in 5,000
Asteroid/comet impact	1 in 6,000
Airplane crash	1 in 20,000
Flood	1 in 30,000
Tornado	1 in 60,000
Venomous bite or sting	1 in 100,000
Fireworks accident	1 in 1,000,000
Food poisoning by botulism	1 in 3,000,000

The problem is that the computation was based on an asteroid/comet about 1.25 km in diameter (a little less than 0.8 of a mi) striking the Earth every 300,000 years or so. If and when it hit, it would have a civilization-destroying potential; that is, it would kill hundreds of millions to billions of people. However, in between, nobody dies from asteroid/comet collisions, while every year numerous people are dying from all the other things on the list. Indeed, remarks Mr. Chapman, "there is no confirmed instance of a single person dying from the impact of an extraterrestrial object" and "the overwhelmingly most likely number of people to die from an asteroid impact in the future is zero." So read the following with the slipperiness of probability in mind.

Bus Metric

• In the United States, buses average 0.4 deaths per billion passenger miles.
• The odds of being killed on a 8-km (5-mi) bus trip is 1 in 500 million.

Dog-Bite Metric	
Dog-Bite Deaths in the United States by Breed	
Breed	**Odds**
Pit bull	37 in 10 million
German shepherd	9 in 10 million
Husky	7 in 10 million
Malamute	6 in 10 million
Doberman	5 in 10 million
Rottweiler	5 in 10 million
Great Dane	4 in 10 million
Saint Bernard	4 in 10 million
General odds of being killed by a dog are 1 in 700,000.	

Household Scary-as-Hell Metric

Dead-Brit Metric	
Cause of Death	**Risks of Death Per Capita Per Year in Great Britain**
Smoking 10 cigarettes a day	1 in 200
Natural causes, 40 years old	1 in 700
Accidents on the road	1 in 10,000
Accidents at home	1 in 10,000
Accidents at work	1 in 50,000
Exposure to nuclear debris	1 in 70,000
All	1 in 80

Roulette-Wheel Metric	
Odds of Even Numbers Coming Up on a Roulette Wheel	
Times	**Odds**
2	1 in 4.2
5	1 in 37
8	1 in 319
10	1 in 1,350
15	1 in 50,000
20	1 in 1.8 million
25	1 in 67 million
28	1 in 578 million
50	1 in 4.4 quadrillion
100	1 in 19^{30}

Odds of Holding Various Poker Hands in the First Five Cards Dealt	
(Number of possible poker hands = 2,598,960)	
Type of Hand	**Odds**
Royal flush	1 in 649,740
Straight flush	1 in 72,193.33
Four of a kind	1 in 4,165
Full house	1 in 694.16
Flush	1 in 508.80
Straight	1 in 254.8
Three of a kind	1 in 47.32
Two pairs	1 in 21.03
One pair	1 in 2.36
No-pair hand	1 in 1.99
Odds of dealing a specific bridge hand from a well-shuffled deck	
1 in 635,013,559,660	
Odds that anyone at a table will get a specific bridge hand	
1 in 158,753,389,900	

Multiple-Birth Metric (North American Scale)	
Type of birth	**Odds**
Twins	1 in 90
Triplets	1 in 9,000
Siamese twins	1 in 100,000
Quadruplets	1 in 900,000
Quintuplets	1 in 85 million

Blood-Type Odds in North America	
Blood type	**Frequency**
O+	37.4
O–	6.6
A+	35.7
A–	6.3
B+	8.5
B–	1.5
AB+	3.4
AB–	0.6

EPHEMERA

1. Odds of two fingerprints being the same, according to the FBI, is 1/64,000,000.

2. Odds you have the subtype h-h blood, the rarest type in the world: 3 in 5.1 billion.

Parts Per

This is not a formal scientific category, but everywhere we look we are besieged by warnings about parts of this per parts of that will lead to the probability in every 100,000 people of something else. How to understand this new way of measuring should be part of the school system. But it isn't, so here goes.

EPHEMERA
1. One person among the roughly 5 billion people on the planet is equal to 200 parts per trillion.
2. A trained tracking dog can follow the sweat scent left by a foot when 400 parts per trillion (4×10^{-11} g or 1.4×10^{-12} oz) are present. This concentration level is 50 to 100 times smaller than the sweat smell left in a fingerprint.
3. Water is considered safe if there are 0.05 parts per million of arsenic or lead in it, and 0.002 parts per million of mercury.
4. A part per trillion is equal to a jigger of Scotch in Lake Superior.

Rough Measures

One part per million equals:

- A second in 277 hours or 11.5 days
- A minute in two years
- The amount of chlorine swimmers can detect in a pool
- Four times the amount of the LSD which the drug's discoverer, Albert Hoffman, first ingested before hallucinating for six hours

One part per billion equals:

- A second in 32 years
- The number of molecules a human nose needs to detect a rose or fuel oil

One part per trillion equals:

- A second in 32,000 years
- One grain of salt in an Olympic-sized swimming pool
- One small gnat's wing in a 100-tonne (98-ton) whale
- The width of one hair in a belt that stretches around the world

One part per quadrillion equals:

- One second in 32,000,000 years
- $\frac{1}{1,000}$ of a gnat's wing in a 100-tonne (98-ton) whale.

Lethal-Dose Metric (with LD_{50} standing for volumes of amounts needed to kill half the test animals)			
Substance	**Animal**	**LD_{50}**	**Lethal parts per body weight**
Acetaminophen (analgesic	mice	0.34 g/kg	3.4/10,000
in Tylenol and Excedrin)			
Acetic acid (vinegar)	rats	3.53 g/kg	3.5/1,000
Arsenic trioxide	rats	0.015 g/kg	1.5/100,000
Aspirin	mice, rats	1.5 g/kg	1.5/1,000
Caffeine	mice	0.13 g/kg	1.3/10,000
(The equivalent of a 70-kg (150-lb) person consuming 70 cups of coffee in one sitting)			
Citric acid	rats (abdominal inject)	0.98 g/kg	9.8/10,000
Ethyl alcohol	rats	13 ml/kg	1/100
		(=10 g/kg)	
Glucose	rabbits intravenous	36 g/kg	36/1,000
Niacin	rats (under skin inject)	5 g/kg	5/1,000
Nicotine	mice	0.23 g/kg	2.3/10,000
Table salt	rats	3.75 g/kg	3.75/1,000
Vitamin B1	mice	8.2 g/kg	8.2/1,000
Most lethal poisons			
Tetanus a		5×10^{-6} mg/kg	5/1,000,000,000,000
Diphtheria		3×10^{-4} mg/kg	3/10,000,000,000
Dioxin		3×10^{-2} mg/kg	3/100,000,000

pH

The most important thing to remember is that each increase of a number represents a tenfold increase in the baseness of a substance.

Food	pH	Nonfood
	0	liquid pool acid (muriatic)
	1	Limeaway bathtub scale remover
	1–3	gastric juices
lime juice	2.3	
lemon juice	2.4	
soft drinks	2.5–3.5	
pickles	2.7	
apples	3.0	
rhubarb	3.1	
grapefruit	3.2	
jelly	3.2–3.4	
cherries or plums	3.4	
sauerkraut	3.5	
orange juice, peaches, pineapple	3.7	
wine	3.0–3.8	
pears	4.2	
tomato juice	4.3	
bananas	4.6	
human urine	4.8–7.5	
carrots	5.2	
baked beans	5.3	
spinach	5.4	
potatoes	5.5	
carbonated water	5.6	
corned beef	5.9	
peas, sardines, evaporated milk	6.0	
pork luncheon meat	6.1	
chicken, butter	6.2	
salmon	6.4	
saliva at rest	6.4–6.9	
shrimp	6.9	
blood and fresh meat and water	near 7.0	
	7.0–7.3	human saliva while eating
	7.8–8.3	sea water
white layer cake	7.5–8.0	
devil's food cake	8–9	
	8.5	baking soda
	around 10	detergents
	10.5	milk of magnesia
	12	household ammonia
	11–14	Drano
	11 or 12	liquid cuticle remover
	13	lye

Acid-Lake Metric	
Acid rain has a pH of between 2.4 and 3.6.	
Seventeen species of snails, shellfish, and crustaceans were found in a Canadian lake which was slowly acidified in the late 1970s and early 1980s.	
At a pH between 7 and 6.2:	17 species remained
Below 6.2:	16 species
At 6.0:	15 species
At 5.9:	11 species
At 5.8:	9 species
At 5.6:	7 species
5.6 to 5.2:	6 species
5.1:	5 species
5.0 to 4.8:	3 species
4.8:	2 species
Below 4.8 to 4:	1 species
Below 4:	Virtually none of the original creatures in the lake, including fish, remains.

Pressure

Pressure is a big force created by lots of pressing but tiny particles. When you force air into your car tires, all the tiny air molecules are racing around madly, colliding with each other and the walls of the tire. Each individual molecule can only exert a small force when it hits the inside wall of your tire, but together, all of them hold up your car! The same is true for liquids. The atmosphere exerts a pressure on us which is compensated for by the internal pressure of our bodies, so we are really not aware of it.

Pressure is expressed as a force per unit area. If a force is uniformly applied to a surface, that surface is under pressure. Normal atmospheric pressure equals one atmosphere (atm). There are a half-dozen scientific measures all used in different disciplines. I won't judge which is best; I will just give you a potpourri to choose from.

Handy Comparisons
1 atm = 1.013 bar = 1.013×10^5 Pa = 760 torr = 14.7 lb/in^2 = 760 mm mercury
1 pascal = 1 newton/m^2 = 1.45×10^{-4} lb/in^2

Pressure Big and Little

	Pressure			
Object or incident	**Pascals**	**atm**	**torr**	**lb/in^2**
Core pressure of the Sun	2.54×10^{16}	2.51×10^{11}	1.9×10^{14}	3.68×10^{12}
Center of the Earth	4×10^{11}	3.95×10^{6}	3.00×10^{9}	5.8×10^{7}
Highest sustained in a lab	1.5×10^{10}	1.48×10^{5}	1.13×10^{8}	2.2×10^{6}
Sustainable by bone	1.63×10^{8}	1.63×10^{3}	1.24×10^{6}	2.40×10^{4}
Deepest ocean trench	1.1×10^{8}	1.09×10^{3}	8.25×10^{5}	1.6×10^{4}
Sustainable by steel	4.13×10^{7}	408	3.10×10^{5}	6.0×10^{3}
Strongest shark bite (long dusky shark)	3.0×10^{7}	296	2.25×10^{5}	4.35×10^{3}
Under the stiletto heels of a 100-kg (220-lb) person	1.42×10^{7}	140	1.1×10^{5}	2.1×10^{3}
In a high-pressure boiler	7.6×10^{6}	75	5.70×10^{4}	1.1×10^{3}
Under the stiletto heels of an average-sized woman	7.6×10^{6}	75	5.70×10^{4}	1.1×10^{3}
Under the foot of an elephant standing on two legs	2.50×10^{6}	24.7	1.88×10^{4}	362
Under the foot of a ballet dancer on one foot	2.50×10^{6}	24.7	1.88×10^{4}	362

Pressure Big and Little *(continued)*				
		Pressure		
Object or incident	**Pascals**	**atm**	**torr**	**lb/in^2**
Under spike heels, dancing	1.1×10^{6}	10.9	8.25×10^{3}	160
Rupturing pressure of veins	5.07×10^{5}	5	3,800	70
Pressure which will kill half the 70-kg (150-lb) men subjected to it	3.44×10^{5}	3.40	2,580	50
Inside a racing bicycle tire	1.79×10^{5}	1.76	1,340	26
Normal blood pressure	1.6×10^{4}	0.158	120	2.32
In a vein in the top of your foot while standing	1.2×10^{4}	0.12	90	1.74
Sinus pressure during acute sinusitis	9,800	9.68×10^{-2}	73.5	1.42
Human jaws	58.8–1,180	5.8×10^{-4}–	0.441–	8.53×10^{-3}–
		1.16×10^{-2}	8.82	0.17
Highest wind pressure in a flute	1,000	9.87×10^{-3}	7.50	0.145
Highest wind pressure in an alto recorder	100–500	9.87×10^{-4}–	0.75–	1.45×10^{-2}–
		4.94×10^{-3}	3.75	7.26×10^{-2}
Ant jaws	35.3–67.6	3.48×10^{-4}–	0.265–	5.12×10^{-3}–
		6.67×10^{-4}	0.507	9.81×10^{-3}
Loudest tolerable sound at 1,000 Hz	30	2.96×10^{-4}	0.225	4.35×10^{-3}
Faintest detectable sound	3×10^{-5}	2.96×10^{-10}	2.25×10^{-7}	4.35×10^{-9}
Sunlight on Earth's surface	3.44×10^{-6}	3.39×10^{-11}	2.58×10^{-8}	4.99×10^{-10}
Lowest detectable sound at 2,750 Hz	2×10^{-8}	1.97×10^{-13}	1.5×10^{-10}	2.90×10^{-12}
Best laboratory vacuum	1×10^{-12}	9.87×10^{-18}	7.50×10^{-15}	1.45×10^{-16}

Soft-Drink Bottle-Pressure Metric			
Average Amount of Pressure Needed to Compress			
Bottle type	**Contents**	**Pressure to Break**	
		kPa	**lb/in²**
2 l plastic (0.53 gal)	Empty	662	96
2 l plastic (0.53 gal)	Cola	2,080	301
1 l plastic (0.26 gal)	Empty	758	110
1 l plastic (0.26 gal)	Soda water	2,320	337
20 oz plastic (0.6 l)	Empty	610	88
20 oz plastic (0.6 l)	Cola	1,470	213
12 oz aluminum (355 ml)	Empty	1,050	152
12 oz aluminum (355 ml)	Cola	3,820	554
12 oz steel (355 ml)	Empty	2,850	414
12 oz steel (355 ml)	Cola	6,410	930
Old-style 6.5-oz Coke bottles (0.19 l)		8,270	1,200

EPHEMERA
1. The highest vacuum achieved by humans is the equivalent of moving basketball-sized molecules from 1 m (1 yd) to 80 km (50 mi) apart.
2. The air pressure required to burst an average 150-mm (5.9-in) condom is between 1.67 and 1.9 kilopascals.
3. Pressure reached in the lightning-strike channel: a megapascal per square meter.
4. Suppose a gangster sprayed Superman with 3-g bullets at a rate of 100 a minute and the speed of the bullets was 500 meters per second. If they bounced back with no loss of speed, the bullets would exert only 5 pascals pressure on Superman's chest.

What about Deadly Low Pressure?

- In near absolute vacuum of outer space 1 to 2 mm mercury (Hg) — which is the equivalent to 133 to 266 pascals per square meter or somewhere between 1⁄380 and 1⁄760 of an atmosphere — a person will lose consciousness in 9 to 11 seconds and probably be dead within 90 seconds.
- A recompression within 60 to 90 seconds to 200 mm/Hg (3.8 lbs per square in), a bit less than 0.25 atmospheres and 26,660 pascals, will revive the person.
- Pressure below 190 mm/Hg (3.7 lbs per square in) is dangerous without proper equipment, below 100 mm/Hg (1.9 lb/in²) is hostile, and below 50 mm/Hg (0.97 lb/in²) would be lethal within 60 to 90 seconds.
- A person can live in air at 190 mm/Hg (3.7 lb/in²) if it is pure oxygen.

• People have survived at 258 mm/Hg (5 lb/in^2) for nearly two months without apparent bad effects.
• Mines 3,100 m (10,000 ft) below sea level experience at least 1,030 mm/Hg (20 lb/in^2) of pressure. That is like 1.5 atmospheres.
• People have survived underwater at a bit more than 2 atmospheres absolute without any trouble. Varying mixtures of oxygen, nitrogen, and helium can sustain sea-level equivalents up to 5.74 atmospheres (84.4 lb/in^2).

Ear-popping Pressure

Ear complaints encountered during changes in barometric pressure

Ascent (mm/Hg)	Complaint	Descent (mm/Hg)
0	No problem	0
+3–5	Fullness in ears	-3–5
+10–15	More fullness, sound less intense	-10–15
+15–30	Fullness, discomfort, tinnitus	-15–30
	Ears pop, air leaves middle ear	
	Desire to clear air, which works	
+30	Increasing pain, tinnitus, dizziness	-30–60
	Severe radiating pain, dizziness, nausea	-60–80
	Voluntary clearing becomes hard or impossible	–100
	Eardrum ruptures	–200+

Radiation

Radiation will drive you nuts, because there are four different kinds of measurements in which various SI and non-SI standards are rather promiscuously used. We have:

Source Activity

Typically defined as curies (Ci), which equal 3.7×10^{10} disintegrations (radioactive decays of the nucleus) per second — approximately the disintegration level found in one gram of radium. The SI unit is the becquerel, which equals one disintegration per second.

A Radiation-Release Scale	
Radiation in curies/becquerel	**Source**
$6 \times 10^{11}/2.2 \times 10^{22}$	20-kilotonne bomb
$10^{10}/3.7 \times 10^{20}$	Nuclear power reactor
$10^{6}/3.7 \times 10^{16}$	Cobalt-60 in high-level industrial irradiation cell
$\sim 10^{3}/3.7 \times 10^{13}$	Cobalt-60 in therapy unit
$5\text{-}50 \times 10^{-6}/18.5\text{-}185 \times 10^{5}$	Iodine-131 for thyroid scan
$\sim 10^{-6}/37{,}000$	Radium-226 in fluorescent paint on wristwatch
$10^{-7}/3{,}700$	Natural potassium-40 in human body
The amount of radiation released at Chernobyl was 50 million curies, or 1.85^{19} becquerel.	

Exposure

Exposure is the amount of radiation reaching a material. It is defined for X-rays and gamma rays in roentgens. One roentgen equals ionization which produces a charge of 2.58×10^{-4} coulombs per kilogram.

A bacteria can withstand 6.5 million roentgens; that is, 10,000 times the dosage which would be fatal to a human.

Absorbed Dose

This is the energy which the radiation imparts to a given amount of the absorbing tissue. It has typically been measured in rads. One rad equals 0.01 joules per kilogram. For comparison, one roentgen exposure to X-rays or gamma rays is approximately equal to one rad of absorbed dose.

The SI unit is the gray. One gray equals 100 rads. For comparison:

• Fertility is affected by as little as 0.15 grays (15 rads), with depressed sperm production lasting for up to a year. Permanent sterility in men will occur at 3.5 grays (350 rads), and a single dose of 1 to 2 grays (100 to 200 rads) will cause temporary sterility. Permanent sterility in women occurs at between 2.5 and 6 grays (250 to 600 rads).

- Average dosages received by the Japanese were 390 milligrays (39 rads) at Hiroshima and 420 milligrays (42 rads) at Nagasaki.
- Skin burns appear at about 3 grays (300 rad). Above 10 grays (1,000 rads), scaling and blistering occur. Hair loss occurs at 3 grays (300 rads).
- Visual impairment generally occurs at about 5 grays (500 rads).

Biological Quantities

These are divided into a number of categories and measures. There is the rem and the millirem. One rem equals one rad of 200-keV X-rays. The SI unit for biological quantities is the sievert. One sievert equals 100 rem.

There is also the Effective Dose Equivalent (EDE), which is the combination of doses given to different organs, which if they were applied uniformly to the whole body would produce the same risk of cancer and genetic damage. A Collective Dose Equivalent (CDE) is the average dose multiplied by the number of people exposed. It has units of man-sieverts.

A Radiation Cause-and-Effect Scale

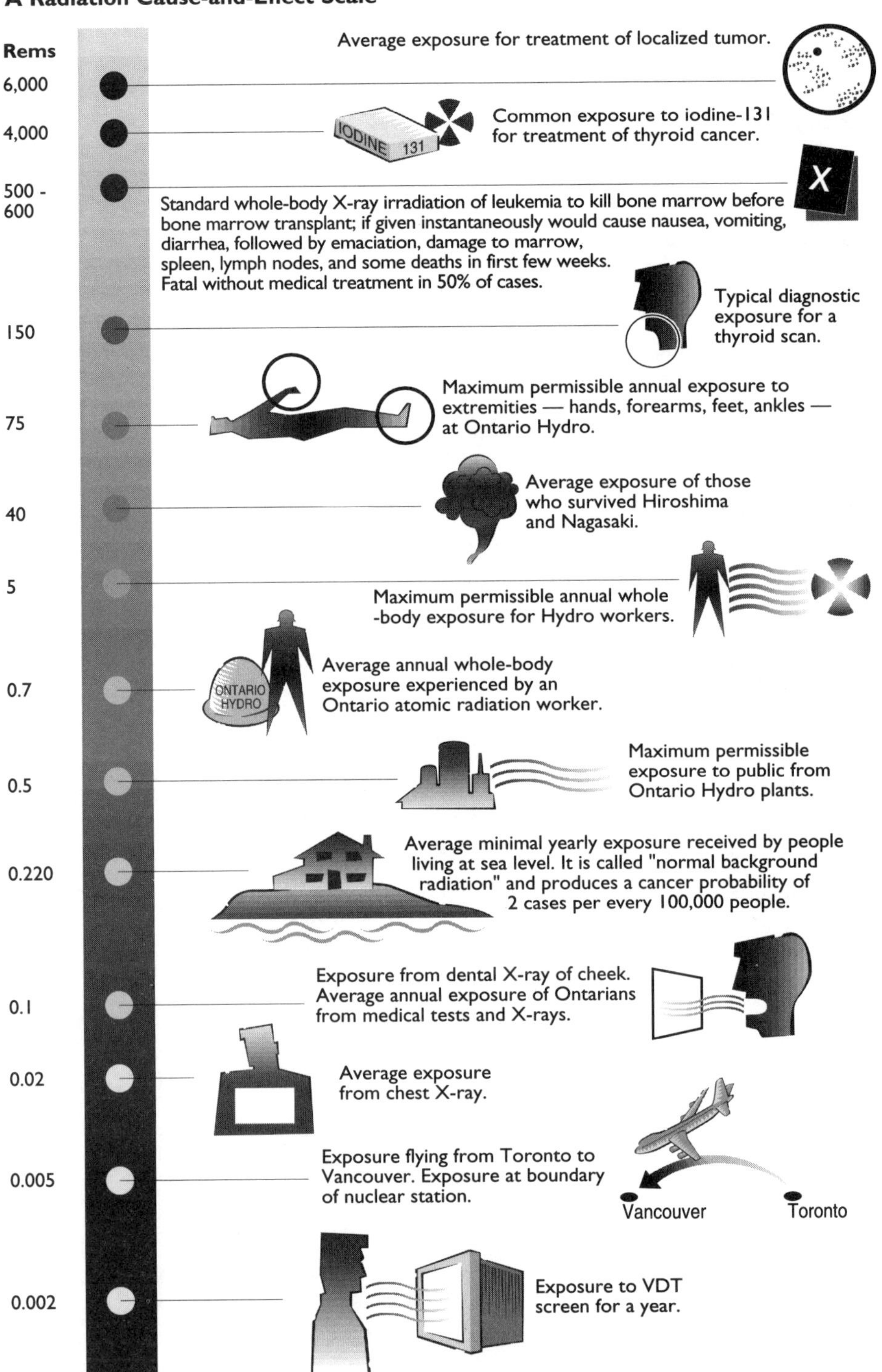

Everyday-Radiation Scale	
Millirems	**Source or location**
	Cosmic rays
41	Sea level
70	Denver (altitude 1,500 m [5,000 ft])
160	Leadville, Colorado (altitude 3,200 m [10,500 ft])
400	6,100 m (20,000 ft)
0.7 per hour	Commercial jet (44 U.S. average)
	Gamma rays from rocks, soil (radium, uranium, thorium, krypton, etc.)
22.8	Atlantic coastal plains
89.7	Colorado front range (40 average from external radionuclides)
26	Average gonadal dose from external radionuclides
16	Internal radionuclides ^{40}K
2	Ra decay products
88 plus or minus 11	Total average gonadal dose
4	Fallout 1970
0.003	Nuclear power
72	Medical diagnostic
1	Medical radiopharmaceutical
0.8	Occupational
2	Misc

Hot Stuff: Radioactivity in Food	
The key thing to remember here is how minute a becquerel is.	
Substance	**Becquerel/gram**
Brazil nuts	6
Cereals	0.25
Teas	0.17
Vegetables	0.003
Meats	0.0015
Fruits	0.0003

EPHEMERA
1. The dose limits should be about 100 mSv in five years and no more than 50 mSv in a year.
2. In Great Britain, average yearly doses ranged from 0.1 mSv in college students to 1.2 in coal miners, 2.0 for aircraft crews, 4.2 for nuclear fuel reprocessing, and 14 for noncoal miners.

The Good Stuff: Nuclear Medicine	
Procedure	**Effective dose equivalents in millisieverts**
Brain scan	5.5
Liver scan	1
Bone scan	4
Thyroid scan	0.8
Indium scan	11.8
Thallium scan	18.4
Chest X-ray	0.05
Skull X-ray	0.15
Abdominal X-ray	1.4
Pelvic CT	2.1
Abdominal CT	2.6
Chest CT	4.8
Barium meal	3.8
Barium enema	7.7
Risk factors for fatal cancer (per sievert) (1990 estimates)	
Bone marrow	0.005
Bladder	0.003
Bone surface	0.0005
Breast	0.0020
Colon	0.0085
Liver	0.0015
Lung	0.0085
Esophagus	0.003
Ovary	0.001
Skin	0.0002
Stomach	0.011
Thyroid	0.0001
Remainder	0.005
Total	0.05

Richter Scale

The things to remember with the so-called Richter scale are three. First, there are about a half-dozen measurements that seismologists take and assign Richter numbers to. You must know what is being measured before you can translate it easily into any equivalent. Second, the Richter scale proper is a measure of the increase in wave size which a seismograph records. On this scale, the numbers increase by a factor of 10. This means that 6 is ten times bigger than 5, which in turn is ten times bigger than 4. However, there is also a separate scale of energy release embedded in the Richter numbers. These increase by 31.6. That is, 6 releases 31.6 times more energy than 5, which itself releases 31.6 more times energy than 4. Finally, there is a damage scale, called a Mercalli scale, which goes up to 12 (or more formally, XII) and which is often coordinated with Richter numbers. This causes people to sometimes say something is a 12 on the Richter scale when they mean an 8 Richter which translates into a 12 Mercalli.

TNT as a Measuring Cup for Earthquakes
One g (0.04 oz) of TNT releases 4.2×10^3 joules
One kg (2.2 lbs) of TNT releases 4.2×10^6 joules
One ton of TNT releases 4.2×10^9 joules.
One kiloton of TNT equals 4.2×10^{12} joules
One megaton 4.2×10^{15} joules.

Richter Scale and the Mercalli Damage Scale Coordinated			
Scale number	**Characteristics**	**Max ground accel m/s²**	**Richter magnitude**
1 instrumental	Only seismograph detects	0.01	less 3.5
2 feeble	Noticed by sensitive people	0.025	3.5
3 slight	Like vibrations of passing truck; felt in upper buildings	0.05	4.2
4 moderate	Felt by people when walking; loose objects and stationary vehicles rock	0.1	4.5
5 rather strong	Generally felt; sleeping people awakened	0.254	4.8
6 strong	Trees sway, suspended objects swing; damage by movements of loose objects	0.5	5.4
7 very strong	General alarm; walls crack and plaster falls	1	6.1
8 destructive	Masonry fissures; chimneys fall; poorly constructed sites damaged	2.5	6.5
9 ruinous	Some houses collapse where ground begins to crack; pipes break open	5	6.9
10 disastrous	Ground cracks badly; many buildings ruined; railway lines bent; landslides in steep slopes	7.5	7.3
11 very disastrous	Few buildings remain standing; bridges destroyed; all services out; great landslides, floods	9.8	8.1
12 catastrophic	Total destruction; thrown into the air; ground rises and falls in waves	>9.8	>8.1

Waiting for the Big One: Earthquake Likelihoods on Original Richter Scale	
Magnitude	**Average number/year above**
8	2
7	20
6	100
5	3,000
4	15,000
3	more than 100,000

Magnitude–Joules	
Magnitude	**Joules**
0	3.1×10^{4}
1	1×10^{6}
2	3.1×10^{7}
3	1×10^{9}
4	3.1×10^{10}
5	1×10^{12}
6	3.1×10^{13}
7	1×10^{15}
8	3.1×10^{16}
9	1×10^{18}
10	3.1×10^{19}

Atomic Bombs as a Measure		
Nuclear explosions	**Joules**	**TNT equivalents**
Novaya Zemlya (Russia 1961)	8×10^{17}	190 Mt
Eniwetok (U.S. 1952)	6×10^{16}	15 Mt
Alamagordo (U.S. 1945)	8×10^{13}	20 kt
Hiroshima	8.4×10^{13}	20 kt
Nagasaki	8.0×10^{13}	20 kt

Little Richter to Biggest

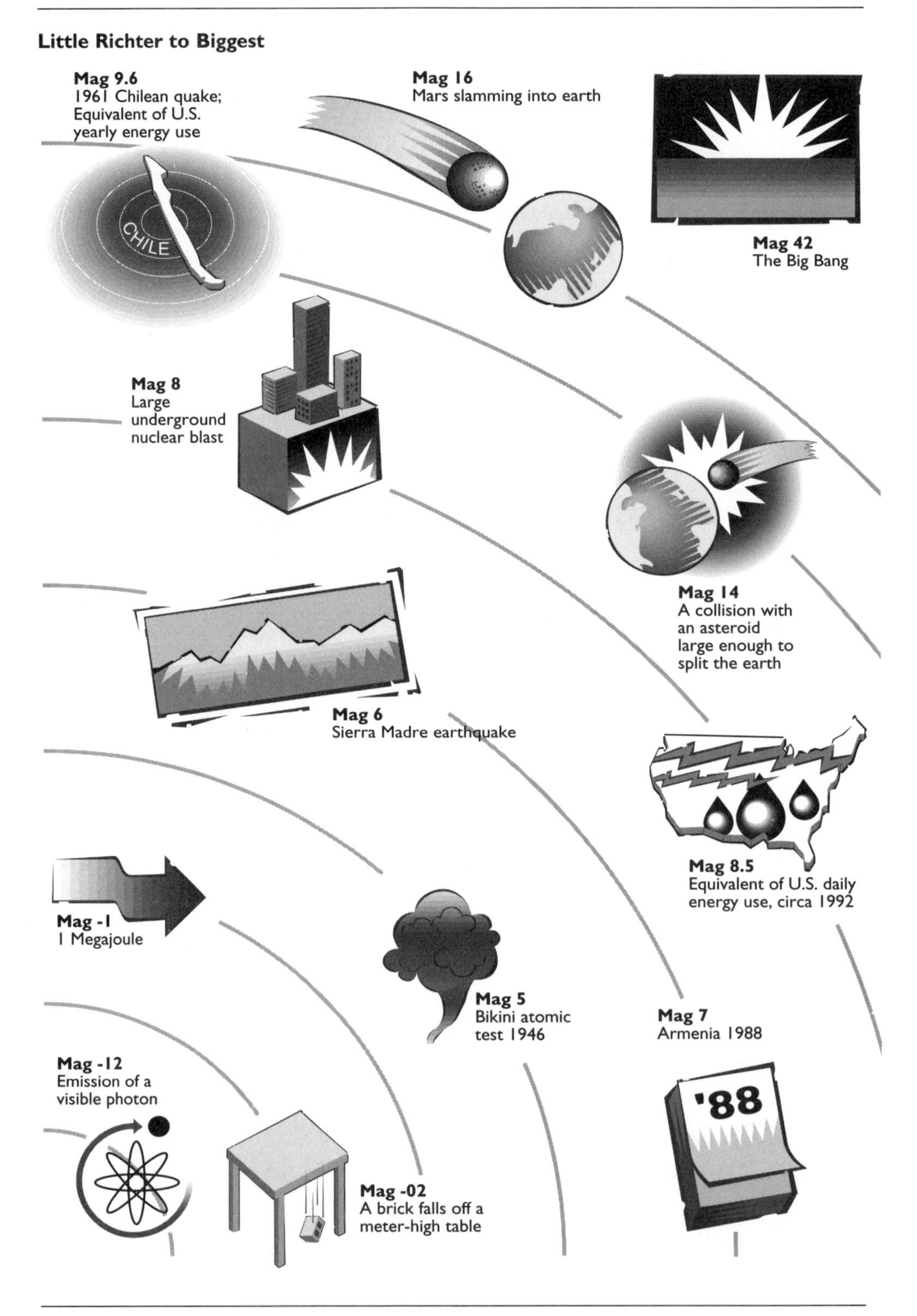

Sound

Once I came across two young girls listening to an echo bounce off their school walls. We started talking about echoes, and I told them about sound being dependent upon an atmosphere to exist. And that meant that whenever *Star Trek* had rocket ships roar off into space they were portraying an impossibility.

"You mean, television lies," said one of the girls.

"Yes," I said. "Isn't it interesting how much we want to hear something which isn't there but which feels as if it should be?"
As *Star Trek* tells us, sound is both something physically real and something biologically perceived.

Sound pitch is measured in hertz. Hertz is, as previously explained, a wave-form measure. One hertz is equal to one wave cycle per second; that is, a wave form which goes from top to bottom, to top again, in one second. A sound frequency is the number of waves which pass a given point in a given time. A high frequency means a lot of waves are going by your starting point. A low frequency means only a few waves are passing by. In terms of our hearing, a high frequency produces sounds of a higher pitch, a low frequency sounds of a lower pitch.

Sounds which are above the normal range of our best hearing — 20,000 hertz — are called "ultrasonic." Those below the lowest sounds we can hear — about 20 hertz — are called "infrasonic."

Sounds are also measured in two other ways. The first is in the intensity or the energy content of the waves. This is measured in watts per square meter (W/m^2). But more commonly this is turned into decibels. Decibels are another logarithmic scale: every time there is an increase of 10 decibels, the energy content of the sound wave has increased tenfold.

The lowest sound people can hear is at 0 decibels, or 10^{-12} watts/m^2. At 60 decibels — the sound of an average conversation — that has turned into 10^{-6} watts/m^2. The loudest sound generated — 210 decibels — produced 10^9 watts/m^2. Sound pressure can also be, like all pressure, measured in pascals.

How Loud Is It?

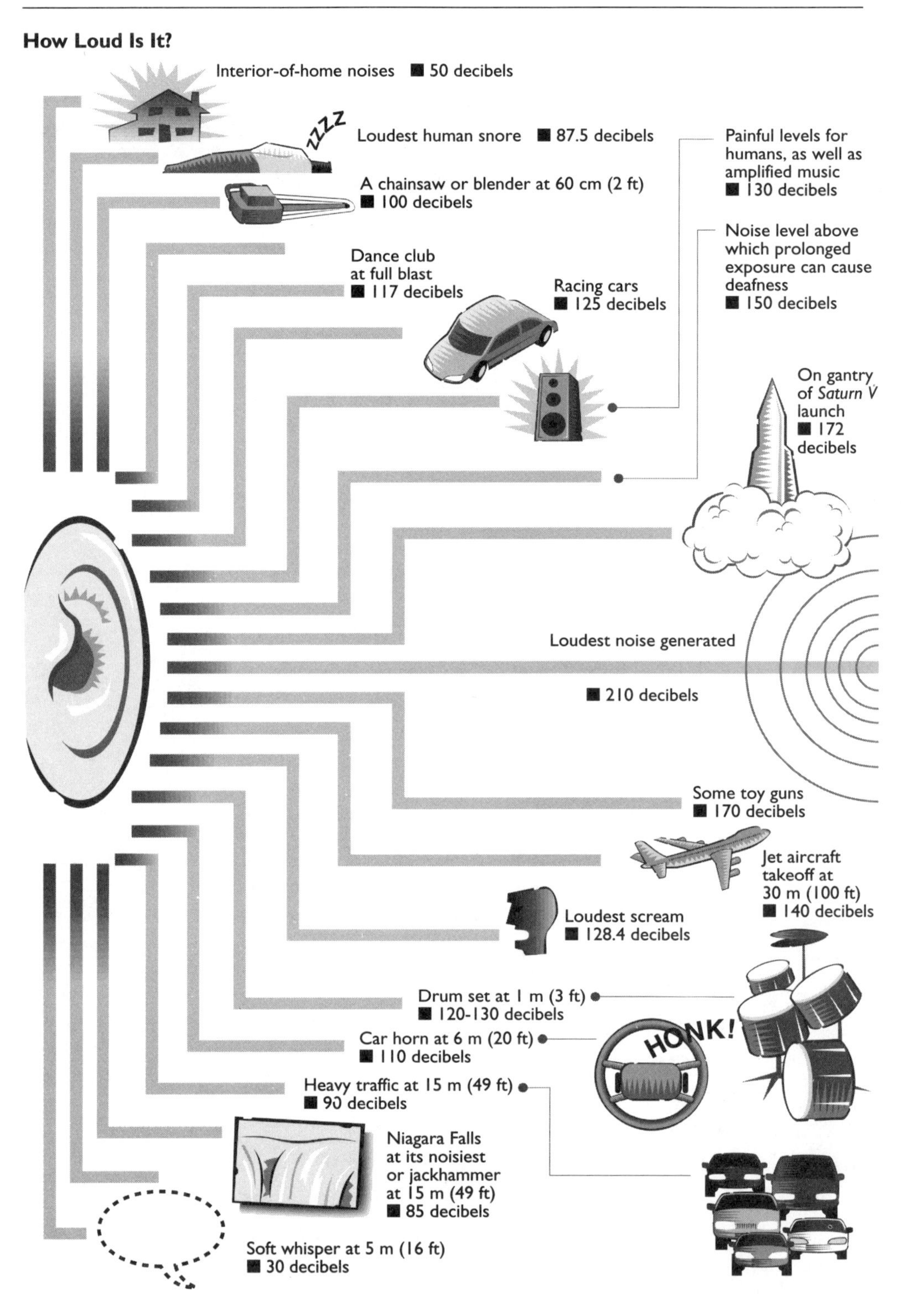

The Human Sound System

- Typical frequencies used in speech are 110 hertz for men, 220 hertz for women, 300 hertz for children.
- The lowest frequency humans can differentiate is 3 to 4 hertz.
- Highest human singing voice has achieved is the E above the piano's top note, 4,350 hertz.
- Lowest is fourth E below middle C, 20.6 hertz.
- Humans can hear 20 to 20,000 hertz.
- Humans can produce 80 to 1,100 hertz.
- Range of singing voices:

    Soprano — from 262 to 1,046 Hz
    Alto — from 196 to 698 Hz
    Tenor — from 147 to 466 Hz
    Baritone — from 110 to 392 Hz
    Bass — from 82.4 to 294 Hz

EPHEMERA
1. Cicadas make a sound of 120 hertz, detectable at 400 m (a quarter of a mile).
2. A 707 DC-8 jet produces a kilowatt per square meter of sound energy, a large truck at 88 km (55 mi) per hour, less than a watt.
3. A 1966 paper by Parrack suggests that ultrasound waves could roast a person to death. Exposure time ranges from 5 minutes at 187 decibels to 40 minutes at 181 decibels.

Musical Physics

Ranges of instruments	Hz
Piano	27.5–4,186
Harp	32.7–3,136
French horn	61.7–698
Guitar	82.4–698
Clarinet	147–1,568
Violin	196–2,093
Trumpet	165–932
Oboe	233–1,397
Flute	262–2,093
Piccolo	587–3,729
When tuned, the four strings on a guitar vibrate at 196, 294, 440, and 660 Hz.	

Animal Voices, Animal Ears		
Species	**Range heard (Hz)**	**Range produced (Hz)**
Dog	15–50,000	452–1,080
Frog	50–10,000	50–8,000
Cat	60–65,000	760–1,520
Grasshopper	100–1,500	7,000–100,000
Dolphin	150–150,000	7,000–120,000
European robin	250–21,000	2,000–13,000
Bat	1,000–120,000	10,000–120,000
Note: Some references say dolphins can make clicks at 170,000 Hz and hear at 280,000 Hz.		

The Power Intensity of Noise

10^{-12} w/m^2: lowest level people can hear

10^{-9} w/m^2: soft whisper at 5 m

10^{-7} w/m^2: interior of home

5 x 10^{-7} w/m^2: average light traffic at 15 m

5 x 10^{-4} w/m^2: jackhammer at 15 m

10^{-3} w/m^2: heavy traffic at 15 m

10^{-2} w/m^2: loud shout at 15 m

10^{-2} w/m^2: jet plane takeoff at 600 m

7 x 10^{-1} w/m^2: discotheque at full blast

1 w/m^2: jet aircraft at 60 m

10 w/m^2: painful level for humans

10^2 w/m^2: jet aircraft takeoff at 30 m

Mechanical Soundings	
	Hz
Woofer (low range)	25–1,000
Squawker	1,000–10,000
Tweeter (high range)	3,000–20,000
Telephone quality speech is between 200 and 3,400 Hz.	

Finally, something straightforward.

Speed

How Fast Is Slow?

Speeder	Speed
Average growth rate of a child between birth and age 18	0.000008 km/h (5×10^{-6} mph)
Growth rate of some lichen	0.0000000001 km/h (0.0000000000625 mph)
Growth rate of bamboo	0.00004 km/h (4×10^{-5} km/h or 2.5×10^{-5} mph)
Speed that a bacteria can move on a kitchen table	0.00016 km/h (0.0001 mph)
Flow rate of some Antarctic glaciers	0.0005 km/h (5×10^{-4} km/h 3.1×10^{-4} mph)
Average speed of Heinz ketchup from the mouth of an upended bottle	0.005 km/h (0.003 mph)
Fastest garden snail's pace	0.008 km/h (0.005 mph)
Speed of a three-toed sloth	0.108–0.158 km/h (0.068–0.098 mph)
Giant tortoise speed	0.27 km/h (0.17 mph)
The speed the body pumps blood to the toes and back	0.6 km/h (0.37 mph)
Fastest centipede's speed	1.8 km/h or 1.1 mph.
Fastest crawling insect: a cockroach	4.64 km/h (2.90 mph)
Speed a crab runs	5.5 km/h (3.4 mph)
Maximum speed the astronauts traveled on the Moon	17.6 km/h (11 mph)

Hair, Fingernail, Breath, Sneeze, and Nerve Metric

• Fingernails grow at 0.05 cm (0.02 in) per week — four times faster than toenails. This translates into a rate of 0.00714 cm (0.0028 in) per day and 0.0003 cm (0.00012 in) per hour.

• Hair grows at a rate of 1.27 cm (0.5 in) per month. However, together all the hairs on our head will roughly grow about 30 m (100 ft) in a day, or a bit more than 11 km (7 mi) in a year. That is, 560 km (350 mi) over a lifetime. By comparison, a merino sheep produces 8,800 km (5,500 mi) of wool in a year.

• The maximum ordinary breathing speed of air passing through the nose equals 3 m (10 ft) per second. This is the equivalent of force 2 on the Beaufort wind scale, i.e., a light breeze.
• Sniffs can occur 32 km (20 mi) per hour, or force 5 on the Beaufort wind scale.
• Sneezes have been recorded to expel particles at 167 km (104 mi) per hour.
• Human nerve impulses travel at 288 km (188 mi) per hour.
• A pricking pain travels at 107 km (67 mi) per hour and a burning pain at 6.5 km (4 mi) per hour.

Fast Machines and Kites	
	Speed
Speed of human-powered flight from Crete to the Island of Santorini	29.7 km/h (18.5 mph)
Top speed for a tank	80 km/h (50 mph)
Top speed of a kite	193 km/h (120 mph)
Fastest steam locomotive	202.8 km/h (126 mph)
Fastest helicopter	400.9 km/h (249.10 mph)
Fastest train	405 km/h (252 mph)
Fastest motorcycle speed	512.7 km/h (318.6 mph)
Fastest speed on water	514.4 km/h (319.6 mph)
Fastest a car has gone, albeit with a jet engine	1,019.4 km/h (633.5 mph)
Fastest airliner (Concorde)	2,333 km/h (1,450 mph)
Fastest jet plane	7,274 km/h (4,520 mph)
Speed of unmanned rocket sled	9,851 km/h (6,121 mph)
Fastest speed of space shuttle	26,715 km/h (16,600 mph)
Fastest speeds humans have achieved on *Apollo 10* trip to moon	39,897 km/h (24,791 mph)

Naturally Fast

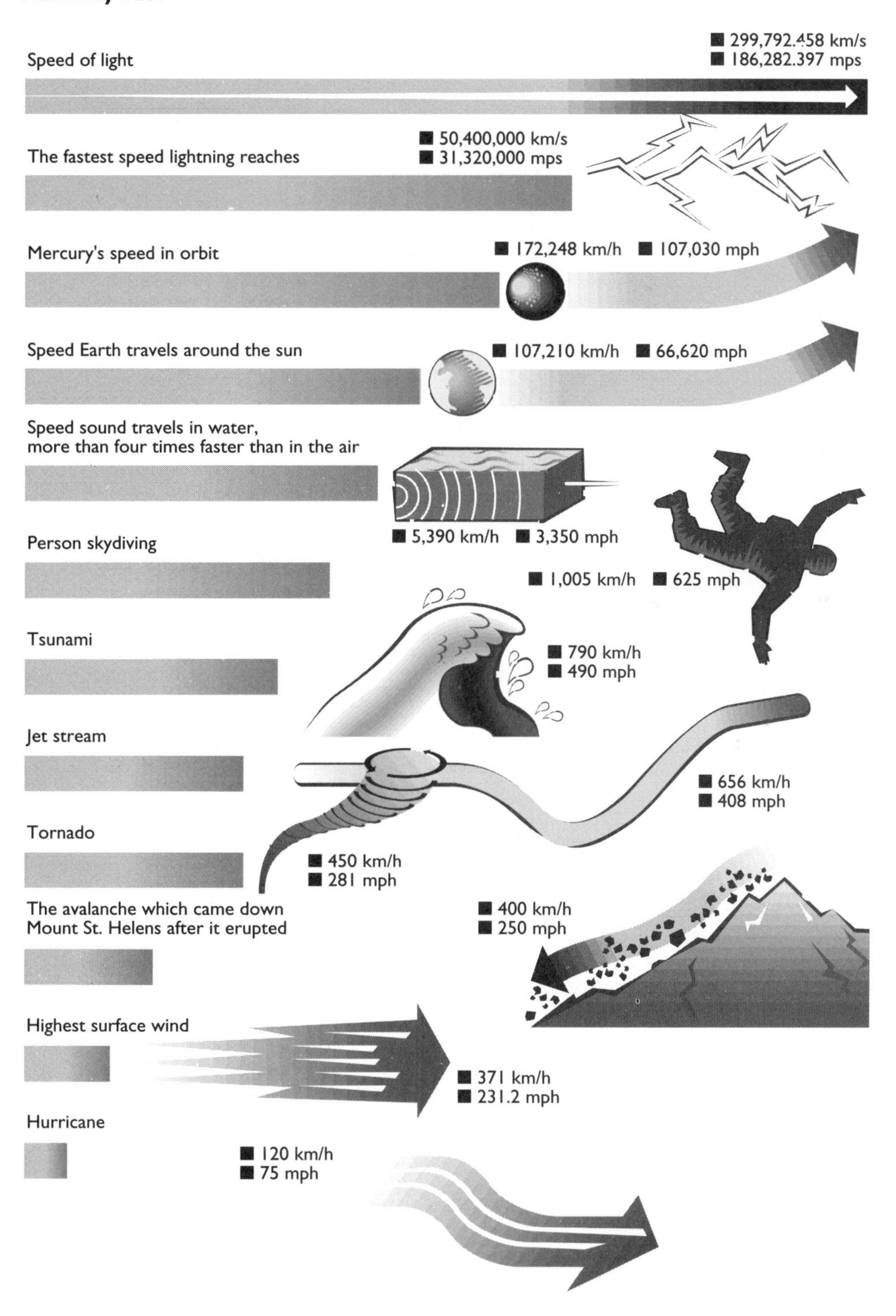

Fast Animals	
	Speed
Fastest spider	16 km/h (10 mph)
Fastest snake speed in a burst	16–19 km/h (10–12 mph)
Speed turtles can travel in water	35 km/h (22 mph)
Speed in short bursts of a sea lion	40 km/h (25 mph)
Jackrabbits' top speed	45 km/h (28 mph)
Speed ostriches can run	64 km/h (40 mph)
Speed greyhound can run	67.14 km/h (41.72 mph)
Fastest a race horse has gone	72.27 km/h (44.91 mph)
Fastest a sailfish can go	110 km/h (68 mph)
Speed cheetah can run in bursts	96–100 km/h (60–63 mph)
Fastest a pigeon has gone with a tail wind	177 km/h (110 mph)
Speed peregrine falcon can dive	350 km/h (217 mph)

Sporting Speed	
	Speed
Fastest Olympic swimmer	8 km/h (5 mph)
Fastest speed recorded in rowing	22 km/h (13.7 mph)
Fastest speed a base runner circled all bases in baseball	29.70 km/h (18.5 mph)
Fastest speed skating	49.38 km/h (30.86 mph)
Fastest a human has run (over 100 m)	54.9 km/h (34.3 mph)
Fastest bike speed without slipstream assistance	105.4 km/h (65.5 mph)
Fastest speed on a luge	137.4 km/h (85.38 mph)
Fastest a baseball has been thrown	162.3 km/h (100.9 mph)
Fastest speed a Ping-Pong ball has been hit	170 km/h (105.6 mph)
Fastest speed on skis	223.7 km/h (130 mph)
Fastest water skiing speed	230.3 km/h (143.1 mph)
Fastest speed a squash ball has been hit	242.6 km/h (150.8 mph)
Fastest bike speed with slipstream aiding	245.1 km/h (152.3 mph)
Fastest tennis serve	263 km/h (163.6 mph)
Speed a golf ball is driven off a tee	273 km/h (170 mph)

Sun-Protection Factor

Sun-protection factors (SPF) are put together on a halving scale. What the numbers represent is percentages of ultraviolet (UV) radiation — or more specifically UV-B, which gets through a material or lotion to your skin. Thus something with an SPF of 2 means that only ½ of the potential UV-B has gotten through to you, and a 3 that only ⅓ has gotten through. Thus, an SPF of 15 means that 6.6 percent of the UV-B has gotten through and 93.4 percent has been blocked. This is usually translated as meaning that one can be out in the sun fifteen times longer than when one's skin is unprotected. Because there is a constant reduction in what remains, increasing SPF numbers reflect a dramatically diminishing blockage return. Thus, an SPF of 30 will block out 96.7 percent of the ultraviolet rays — only 3 percent more than the SPF 15.

Temperature

Handy Conversion Guide

The Kelvin scale starts at absolute zero — –273.15°C or –459.67°F.

The Reaumur scale has 0 as the freezing point of water and 80 as the boiling point.

To convert Fahrenheit to Celsius, subtract 32 and then multiply by 5/9.

To convert Celsius to Fahrenheit, multiply the C temperature by 9/5 and then add 32.

EPHEMERA
1. 40 – 60°K (–213 – 233°C): The temperature the Earth would become if the Sun were turned off (heat still coming from inside).
2. –143°C (–224.4°F): The lowest temperature in our atmosphere.
3. 306°C (583°F): Temperatures bacteria can thrive at.
4. Mercury thermometer would freeze at –39°C (–38°F), while a liquid alcohol thermometer would work until –114°C (–173°F).
5. It would take about five years for the mean temperature of the ocean to fall a mean of one Celsius degree after the sun was turned off.
6. If the ground is colder than –23°C (–10°F), the ice between your feet can't melt but will instead snap.
7. Revelation 21:8 says there are lakes of sulfur in Hell. On the Earth,

Hot, Hot, Hot

- Above 45°C (113°F) on any area of skin an average person will feel pain.
- A knee can tolerate 47°C (117°F) on metal surfaces for up to half a minute without feeling pain; a hand for only 10 to 15 seconds.
- Fingertips in leather gloves can tolerate metal surfaces heated to 65–71°C (150–160°F) for between 7.3 and 13 seconds. It can be twice this hot on the palm.
- Palms can endure 149°C (300°F) in an Arctic mitten for 18.7 seconds, and for 37 seconds in a leather glove.
- In a pigskin heat glove, a palm can endure 427°C (800°F) for 18.5 seconds.

Warm to Cold		
Normal body temp is 37°C (98.6°F)		
Average skin temperature	**Typical sensation**	
above 35°C (95°F)	Unpleasantly warm	
34°C (93.2°F)	Uncomfortably warm	
below 31°C (87.8°F)	Uncomfortably cold	
30°C (86°F)	Shiveringly cold	
29°C (84°F)	Extremely cold	
When hands reach	**When feet reach**	**They feel**
20°C (68°F)	23°C (73.4°F)	Uncomfortably cold
15°C (59°F)	18°C (64.4°F)	Extremely cold
10°C (50°F)	15°C (59°F)	Painful and numb

It's Colder Than It Looks

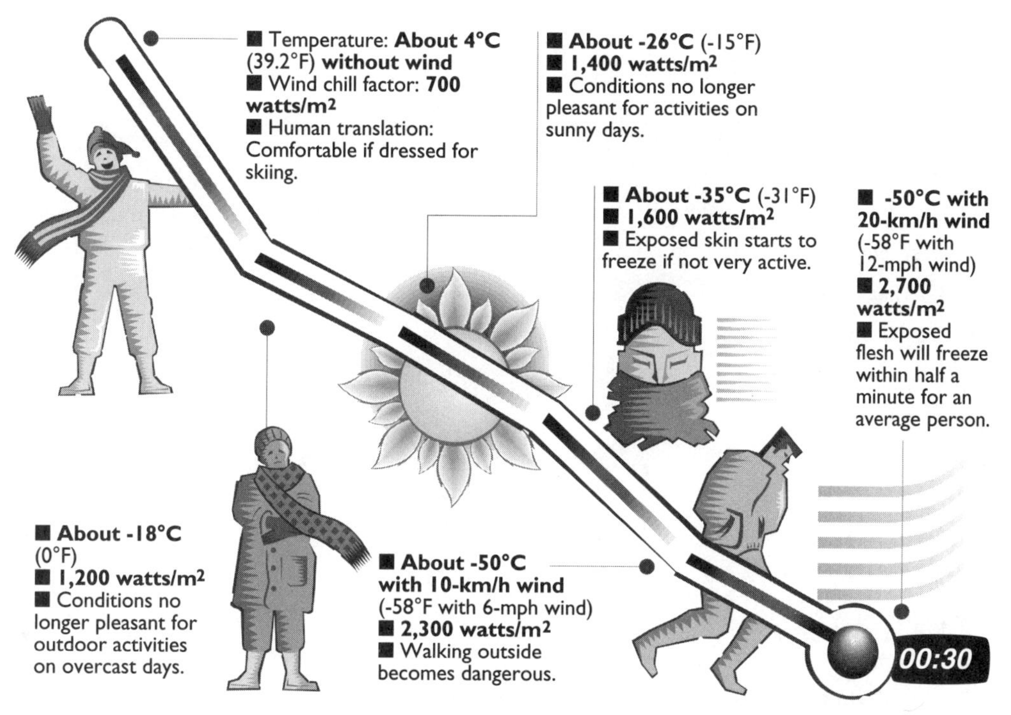

Too Hot to Work

The Humidex Scale

20 to 29 is comfortable.

30 to 39 causes some discomfort.

40 to 45 is uncomfortable for everyone.

Over 46, many types of work must be stopped.

46 = a temp of 32°C (90°F) and a relative humidity of 75, or a temp of 30°C (86°F) and a relative humidity of 90.

Miscellaneous Feats by Humans Running Hot and Cold

- People have walked across coals heated to 841°C (1,546°F).
- People have endured temperatures of 204°C (400°F) naked and 260°C (500°F) with clothes on.
- Saunas at 140°C (284°F) can be enjoyed comfortably.
- Hottest body temperature a person has survived is 46.5°C (115.7°F).
- Coldest body temperature a person has been known to have and survive is 16°C (60.8°F).

sulfur turns to gas at 444.6°C (832°F). Therefore, if the pressure in Hell is one atmosphere, the temperature of Hell cannot be higher than 446.6°C (832°F). However, under extreme pressure, the temperature could reach 1,040°C (1,904°F).
8. To reach a voluntary product standard, refillable soft-drink bottles should be able to withstand a temperature difference of 42°C (75°F) without cracking. Typically, this means they are taken from a bath of hot water at about 63°C (about 145°F) and within fifteen seconds are dumped into a cold-water bath of 21°C (70°F).

Kelvin, Celsius, and Fahrenheit Compared			
Phenomenon	**°K**	**°C**	**°F**
Coldest temperature possible	0	–273.15	–459.67
Coldest temperature obtained in lab	2×10^{-9}	–272.99999	–459.399999
Outer space	3	–270	–457
Liquid helium	4	–269	–452
Melting point of nitrogen	67	–209	–344
Gasoline freezes	123	–150	–238
Lowest surface temperature	183.8	–89.2	–128.6
Dry ice (CO_2 freezes)	195	–78	–108
Water freezes	273.15	0	32
Hibernating squirrel	275	2	35.6
Butter melts	304	30.6	87
Body temperature in man	310	37	98.6
Body temperature in birds	315	42	107.6
Earth's highest temp. in shade	331	58	136.4
Water boils	373.15	100	212
Venus	750	480	890
Fireplace fire	1,100	827	1,520
Gold melts	1,336	1,063	1,945
Gas flame (stove)	1,900	1,627	2,960
Molten lava	2,000	1,730	3,150
Surface of Sun	6,000	5,727	10,340
Center of Earth	16,000	~15,700	~28,800
Temperature of lightning	30,000	~30,000	~54,000
Center of Sun	100,000,000	~100,000,000	~180,000,000
At center of H-bomb	300,000,000 to 400,000,000 C (or K)		540,000,000–720,000,000

Time

The great thing about time is it is in units that everyone understands and uses. Except, of course, it isn't clear why things are divided into units of 60. The answer is that while 10s are divisible only by 1, 2, and 5, 60 is divisible by 1, 2, 3, 4, 5, 6, 10, 12, 15, 20, and 30; that is, it turns into lots of little Time Pieces, and that appealed to the Sumerians, who first devised the 60-unit-long system. It also appeals to me.

A Time Breakdown
A day = 86,400 seconds = 1,440 minutes.
A week = 604,800 seconds = 10,800 minutes = 168 hours.
A month = 2,592,000 seconds = 43,200 minutes = 720 hours = 30 days = 4.28571 weeks.
A year = 31,556,925.51 seconds = 525,948.75 minutes = 21,914.531 hours = 365.2419 days = 52.17746 weeks = 1.03176 common lunar years = 1.03069 lunar astronomical years = 0.999961 sidereal (sun) years.
It takes 230,000,000 years for our galaxy to complete one rotation.

Milliseconds Turned into Sports Impact Times

Impact	Time in seconds
Golf ball hit by driver	0.001
Baseball off tee	0.0013
Softball off tee	0.0035
Tennis (forehand)	0.005
Football (kick)	0.008
Handball (serve)	0.0125
Soccer ball (header)	0.023

Events in Seconds

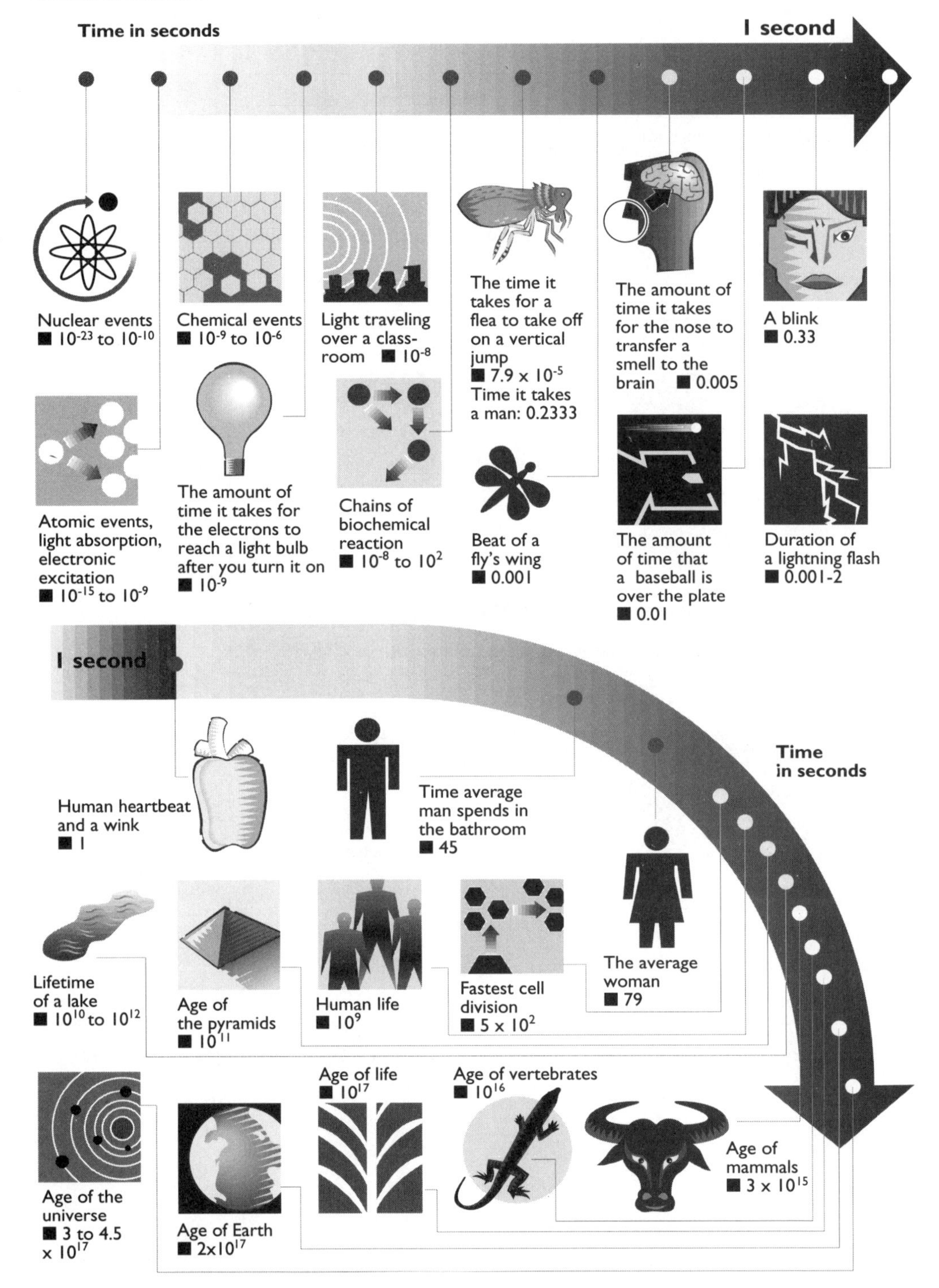

Earth History as a Year

If the history of the Earth were a year, then:

- life appeared at the end of March
- the first fish appeared on November 27
- the first dinosaurs appeared on December 15 and died out on December 26
- mammals first appeared about the same time
- hominids appeared at 7:40 p.m. December 31
- Columbus discovered America 4 seconds before midnight
- the lifespan of a person living to be 120 would be 84/100 of a second

A Minute in Terms of Heartbeats of an Animal at Rest

	Beats-1 minute
Shrew and blue-throated hummingbird	1,200
Canary	1,000
Mouse	650
Elephant	500
Porcupine	280
Hamster	280
Chicken	200
Chihuahua	120
House cat	110
Saint Bernard	80
Human	60–80
Giraffe	60
Tasmanian devil	54–66
Kangaroo and tiger	40–50
Beluga whale	15–16
Gray whale	8

Humans also blink about 15 times a minute.

Age Is Not Ageless

Oldest person: Shigeychico Izumi, 120

Oldest animals: Tortoise, 152; Clam, 220

Oldest plant: Bristlecone pine, 5,100

EPHEMERA

1. Edison's first light bulb lasted 40 hours; a modern one can last 2,500 hours.

2. Estimated amount of time smoking one cigarette takes off an average person's life: 1½ minutes.

3. We dream for 2 hours a night.

4. The time it took the first lighter-than-aircraft the *Graf Zeppelin*, to travel around the world: 21 days, 7 hours, and 34 minutes.

5. Number of chews per minute while eating range from 49 to 120, with average being 70 to 80.

6. Time that it took for first *Sputnik* to go around the world: 96.2 minutes.

7. A dragonfly flaps its wings 20 to 40 times a second, bees and houseflies 200 times, some mosquitoes 600 times, and a tiny gnat 1,000 times.

8. The amount of time it took to complete the great pyramid at Giza in Egypt: 33,000 man-years.

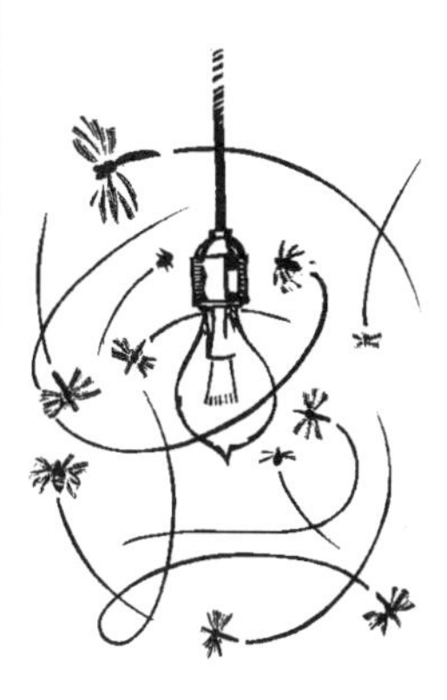

Volume

Volume seems easy, but there are several different kinds of measures at work. There can be volume, which is generally measured in cubic meters and cubic feet or cubic liters, and there is liquid capacity, which is measured in liters or quarts or gallons. In Imperial measure, there are also barrels and bushels and pints and cups and spoons and drops and god-knows-what. If ever there is a place where the intrinsic advantage of metric can show itself to the ordinary person, it is here.

Kitchen Volumes		
Measuring instrument	**Liters or ml**	**Quarts or oz**
Teaspoon	4.93 ml	0.17 oz
Tablespoon	14.79 ml	0.50 oz
Cup	236.64 ml	8.00 oz
Wineglass	300 ml	10.1 oz
Coke can	355 ml	12.0 oz
Wine bottle	750 ml	25.3 oz or 0.80 quarts
Thermos	1 liter	1.1 quarts
Mixing bowl	2 liters	2.1 quarts
Small saucepan	1.4 liters	1.5 quarts
Medium saucepan	1.9 liters	2 quarts
Large saucepan	4.7 liters	5 quarts
Plastic bucket	9 liters	9.5 quarts

Human-Heart Volumetric

This sounds easy, but in fact because we have entered the arena of the human, we are also now in the arena of the variable. Normal ranges of blood volume pumped by the heart when men and women are lumped together are probably between 4 liters (1 U.S. gallon) and 7 liters (2 U.S. gallons) a second. A trained marathoner might pump out as much as 30 liters (8 U.S. gallons) per second. I will go for a middle ground and assume that a figure of about 5 liters (1.3 U.S. gallons) is reasonable. And so all the following figures are based on that theoretical person.

Volume Pumped by Heart		
Time	**Liters**	**U.S. gallons**
One minute	300	79
One hour	18,000	4,760
One day	432,000	114,000
One year	157,680,000	41,627,000
75 years	11,826,000,000	3,122,000,000
By way of comparison, that means that in a normal lifetime the human heart pumps enough blood to fill up the world's largest oil tanker more than 46 times.		

Outhouse Volumetric		
Service life of a pit outhouse for five people		
Years	**Cubic meters**	**Cubic feet**
4 years	1.16	41
8 years	2.29	81
15 years	4.24	150

Water-Usage Volumetric		
Machine	**Liters**	**U.S. gallons**
Flush toilet	20	5.28
Shower uses	25 per minute	6.6 per minute
Bathtub	holds about 130	34.3
Dishwasher	65	17.16
Clothes washer	about 230	60.72
Garden hose	1,500 or more an hour	396 or more an hour
Dripping tap	30 to 100 a day	7.92 to 26.4

Body Volumetric		
Lungs	**Lung volume**	
	Liters	**Quarts**
Whale	100	105.65
Cow	10	10.565
25-year-old man	6.44	6.8
25-year-old woman	4.16	4.4
Cat	0.5	0.528
Human sinuses	0.05	0.0528
Rat	0.01	0.0105
Mouse	0.001	0.00105

Champagne Volumetric		
Bottle size	**Quarts**	**Liters**
Wine bottle	0.80	0.76
Magnum = 25 wine bottles	2.0	1.9
Jeroboam = 1.6 magnums	3.2	3.0
Rehoboam = 3 magnums	6	5.7
Methuselah = 4 magnums	8	7.6
Salmanazar = 6 magnums	12	11.4
Balthazar = 8 magnums	16	15.1
Nebuchadnezzar = 10 magnums	20	18.9

Common Measures

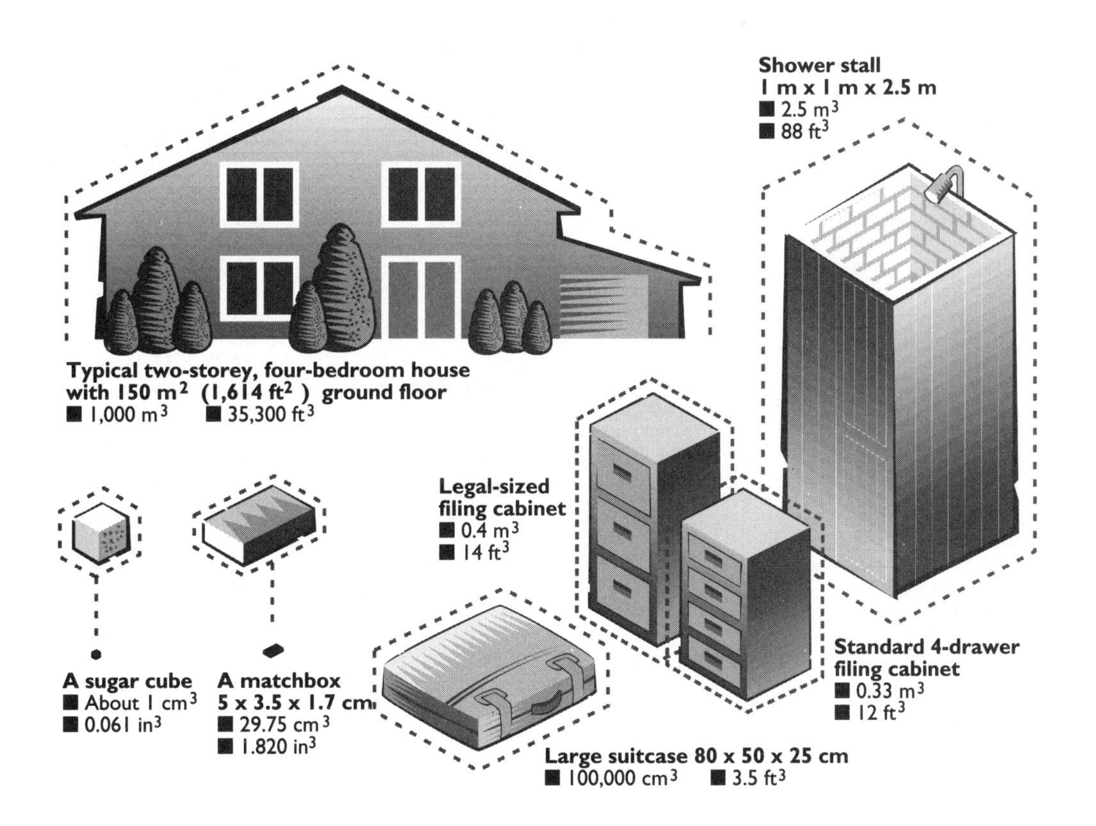

EPHEMERA

1. A newly laid egg takes up 6 liters (1.58 gallons) of oxygen and gives off 4.5 liters (1.19 gallons) of carbon dioxide and 11 liters (2.9 gallons) of water vapor during the 21 days it takes to hatch.
2. An acre (4,000 m²) of rainforest in Java moves an estimated 5.7 million liters (1.5 million U.S. gallons) of water up its stalks.
3. A standard 150-mm (5.9-in) long condom, when inflated, can hold about 30 liters (1830 cubic in) of air before bursting. If it is 100 mm (3.9 in) long, it will hold around 1.9 liters (116 cubic in).
4. Minimal difference in capacity allowed in glass soft-drink bottles in the United States is 0.005 liters (0.16 fluid oz) in a 0.3-liter (10-fluid-oz) bottle to 0.011 liters (0.36 fluid oz) in a 1-liter (33.82-fluid-oz) bottle.
5. The volume of the Earth would be equivalent to 1,083,208,840,000,000,000,000,000,000 or about 10^{27} sugar cubes.
6. Filter feeders must move 907 kg (one ton) of water to get 28 g (one oz) of food.
7. A camel can drink 114 liters (30 U.S. gallons) in 10 minutes, and 189 liters (50 gallons) over several hours.
8. The body's 3 million sweat glands will pump out 2 liters (a bit more than 2 quarts) of perspiration on a summer's day, a number which can go to 9.5 liters (2.5 gallons) in a hot, deserty area.
9. Quetzlcoatl, the largest pyramid in the world, contains 3,333,500 m³ (4,360,000 cubic yards) of material. The pyramid at Giza contains 2,568,900 m³ (3,360,000 cubic yards).
10. Grand Coulee Dam contains 8,027,800 m³ (10,500,000 cubic yards) of concrete.
11. Total volume of the world's oceans is 1,349,900.000 km³ or 324,000,000 cubic mi.
12. The moon's volume is 21,900,000,000 km³ (5,260,000,000,000 cubic mi).
13. The flow of the Nile in a year is 83,333 million m³ (3.3 million cubic ft).
14. The Amazon discharges an average of 120,000 m³ (4,200,000 cubic ft) per second into the Atlantic.
15. The Earth's volume is 1,083,208,840,000 km³ (260,000,000,000 cubic mi), nearly 50 times that of the Moon.
16. The size of the jars of water Jesus turned into wine is 76–114 liters (20–30 gallons).
17. One 10.6–12-m tree (35–40-ft) produces a stack of newspapers 1.2 m (4 ft) thick. *The Globe and Mail* measures about 13½ in x 11¾ in (folded) or 1,023 cm² (158.63 square in), to give a stack of newspapers about 0.123 m³ (4.41 cubic ft) in volume.
18. Nine trillion medium-sized bacteria will fill a box the size of a package of chewing-gum sticks.

Winds

Hurricanes							
Saffir/Simpson Hurricane Scale ranges							
Scale number (category)	**Barometric pressure (inches of mercury)**	**Millimeters**	**Winds mph**	**km/h**	**Surge feet**	**meters**	**Damage***
1	>28.94	>735	74–95	120–53	4–5	1.2–1.5	Minimal
2	28.50–28.91	724–34	96–110	154–77	6–8	1.8–2.4	Moderate
3	27.91–28.47	709–23	111–30	178–209	9–12	2.7–3.7	Extensive
4	27.17–27.88	690–709	131–55	210–49	13–18	4.0–5.5	Extreme
5	<27.17	<690	>155	>249	>18	>5.5	Catastrophic

***Translation of the Damage Scale**

Minimal: No real damage to buildings; some flooding; tree, shrubbery, mobile-home damage.

Moderate: Some roof, window, door damage; vegetation damage considerable; low-lying and coastal areas flooded several hours before center of storm arrives; small craft can break mooring in protected areas.

Extensive: Some structural damage to small or residential buildings; mobile homes destroyed; flooding near coasts destroys structures and floods homes 1.5 m (5 ft) above sea level.

Extreme: Roof, window, and door damage; major damage to lower floors of structures near the shore, and some roof-failure on small residences; complete beach erosion; flooding of terrain 3 m (10 ft) above sea level as far as 10 km (6 mi) inland requires massive evacuations.

Catastrophic: Complete failure of main buildings; small utility buildings blown away; major damage to lower floors of all structures 6 m (19 ft) above sea level located within 450 m (500 yds) of the shoreline; massive evacuations 8 to 16 km (5 to 10 mi) from shore.

It's Not a Tornado but It's Still Uncomfortable		
A Draftiness Scale		
Meters/second	**Miles/hour**	**Effect on humans**
Up to 0.25	0.56	Unnoticed
0.25 to .50	0.56 to 1.12	Pleasant
0.50 to 1.00	1.12 to 2.24	Awareness of air movement
1.00 to 1.50	2.24 to 3.36	Drafty
Above 1.50	Above 3.36	Annoyingly drafty

Conclusion: A silly end is better than no end

Having finished this book, I still don't know if all measures are intrinsically silly, or if there is something about measuring which brings out the silly in people. I do know scientists become exhausted with their never-ending mensuration and develop various ways of making fun of it, and themselves. Consider the following effort to take metric's prefixes and turn them into measure puns.

10^{-15} bismols = a femto-bismol

10^{-12} boos = a picoboo

1 boo^2 = a boo-boo

10^{-18} boys = an attaboy

10^{12} bulls = a terabull

10^{10} = a decacard

10^{-9} goats = a nanogoat

2 gorics = a paregoric

10^{9} los = a gigalo

10^{-1} mates = a decimate

10^{-2} mentals = a centimental

10^{-2} pedes = a centipede

10^{6} phones = a megaphone

10^{-6} phones = a microphone

10^{12} pins = a terapin[1]

50 figanewtons = 1 boxanewtons

1 teranewton = 1 bigapileafignewtons

1 microcentury = the average length of a school hour[2]

Added to which there are groupings of an imprecise number which connote a bunch, a mess, a multitude. These are a little in-groupy, but still silly enough to include.

Scientists come in:

A pile of nuclear physicists

A grid of electrical engineers

A set of pure mathematicians

A field of theoretical physicists

An amalgamation of metallurgists

A line of spectroscopists
A coagulation of colloid chemists
A galaxy of cosmologists
A cloud of theoretical meteorologists
A litter of geneticists
A knot of nautical engineers
A labyrinth of communications engineers
A stack of librarians
A complex of psychologists
A wing of ornithologists
A bath of fermentation chemists
A colony of bacteriologists[3]

Some measure jokes have achieved mystical proportions. Consider the Smoot. In the 1950s, a group of Lambda Chi Alpha pledges at the Massachusetts Institute of Technology were told to measure the Massachusetts Avenue Bridge over the Charles River using Oliver R. Smoot Jr. as their measuring rod. Flipping him end to end, they found the bridge to be 364.4 Smoots, plus one ear.

And finally, what silliness people would get into if they would be allowed to make their own measuring systems. In 1984, *Harper's* magazine asked a whole group of magazines what kind of Eden they would construct. Some included weights and measures. The editors of *The National Lampoon* said: "In an effort to reinstate babies to their proper place as the basis of all civilization, the first baby born each year would become the standard for weights and measures, 'That would weigh about one and half Debbies,' we would say, or 'It's about seventeen Maxes long.'"[4]

In the same issue, *Forbes* magazine editor Sheldon Zalaznick said his magazine favored an Eden with everything in metric "but lots of slang terms for various portions, e.g., a gaffe of gin (10 milliliters), a fiasco of gasoline (fill 'er up)."[5]

My only response to all this is: Me^2 Sheldon, me^4 and me 5 x 5 x 5 x 5 x 5.

Appendix

Prefixes

The following prefixes, in combination with the basic unit names, provide the multiples and submultiples in the International System. For example, the unit name "meter," with the prefix "kilo" added, produces "kilometer," meaning "1,000 meters."

Prefix	Symbol	Multiples	Equivalent
exa	E	10^{18}	quintillionfold
peta	P	10^{15}	quadrillionfold
tera	T	10^{12}	trillionfold
giga	G	10^{9}	billionfold
mega	M	10^{6}	millionfold
kilo	k	10^{3}	thousandfold
hecto	h	10^{2}	hundredfold
deka	da	10	tenfold
deci	d	10^{-1}	tenth part
centi	c	10^{-2}	hundredth part
milli	m	10^{-3}	thousandth part
micro	µ	10^{-6}	millionth part
nano	n	10^{-9}	billionth part
pico	p	10^{-12}	trillionth part
femto	f	10^{-15}	quadrillionth part
atto	a	10^{-18}	quintillionth part

Tables of Metric Weights and Measures

Linear Measure

10 millimeters (mm)	= 1 centimeter (cm)
10 centimeters	= 1 decimeter (dm) = 100 millimeters
10 decimeters	= 1 meter (m) = 1,000 millimeters
10 meters	= 1 dekameter (dam)
10 dekameters	= 1 hectometer (hm) = 100 meters
10 hectometers	= 1 kilometer (km) = 1,000 meters

Area Measure

100 square millimeters (mm^2)	= 1 square centimeter (cm^2)
10,000 square centimeters	= 1 square meter (m^2) = 1,000,000 square millimeters
100 square meters	= 1 are (a)
100 ares	= 1 hectare (ha) = 10,000 square meters
100 hectares	= 1 square kilometer (km^2) = 1,000,000 square meters

Fluid Volume Measure

10 milliliters (mL)	= 1 centiliter (cL)
10 centiliters	= 1 deciliter (dL) = 100 milliliters
10 deciliters	= 1 liter (L) = 1,000 milliliters
10 liters	= 1 dekaliter (daL)
10 dekaliters	= 1 hectoliter (hL) = 100 liters
10 hectoliters	= 1 kiloliter (kL) = 1,000 liters

Cubic Measure

1,000 cubic millimeters (mm^3)	= 1 cubic centimeter (cm^3)
1,000 cubic centimeters	= 1 cubic decimeter (dm^3) = 1,000,000 cubic millimeters
1,000 cubic decimeters	= 1 cubic meter (m^3) = 1 stere = 1,000,000 cubic centimeters = 1,000,000,000 cubic millimeters

Weight

10 milligrams (mg)	= 1 centigram (cg)
10 centigrams	= 1 decigram (dg) = 100 milligrams
10 decigrams	= 1 gram (g) = 1,000 milligrams
10 grams	= 1 dekagram (dag)
10 dekagrams	= 1 hectogram (hg) = 100 grams
10 hectograms	= 1 kilogram (kg) = 1,000 grams
1,000 kilograms	= 1 metric ton (t)

Table of U.S. Customary Weights and Measures

Linear Measure

12 inches (in)	= 1 foot (ft)
3 feet	= 1 yard (yd)
5½ yards	= 1 rod (rd), pole, or perch (16½ feet)
40 rods	= 1 furlong (fur) = 220 yards = 660 feet
8 furlongs	= 1 statute mile (mi) = 1,760 yards = 5,280 feet
3 miles	= 1 league = 5,280 yards = 15,840 feet
6076.11549 feet	= 1 International Nautical Mile

Liquid Measure

When necessary to distinguish the liquid pint or quart from the dry pint or quart, the word "liquid" or the abbreviation "liq" should be used in combination with the name or abbreviation of the liquid unit.

4 gills	= 1 pint (pt) = 28.875 cubic inches
2 pints	= 1 quart (qt) = 57.75 cubic inches
4 quarts	= 1 gallon (gal) = 231 cubic inches = 8 pints = 32 gills

Area Measure

144 square inches	= 1 square foot (ft^2)
9 square feet	= 1 square yard (yd^2) = 1,296 square inches
30¼ square yards	= 1 square rod (rd^2) = 272¼ square feet
160 square rods	= 1 acre = 4,840 square yards = 43,560 square feet
640 acres	= 1 square mile (mi^2)
1 mile square	= 1 section (of land)
6 miles square	= 1 township = 36 sections = 36 square miles

Cubic Measure

1 cubic foot (ft^3)	= 1,728 cubic inches (in^3)
27 cubic feet	= 1 cubic yard (yd^3)

Gunter's or Surveyors' Chain Measure

7.92 inches (in)	= 1 link
100 links	= 1 chain (ch) = 4 rods = 66 feet
80 chains	= 1 survey mile (mi) = 320 rods = 5,280 feet

Troy Weight

24 grains	= 1 pennyweight (dwt)
20 pennyweights	= 1 ounce troy (oz t) = 480 grains
12 ounces troy	= 1 pound troy (lb t) = 240 pennyweights = 5,760 grains

Dry Measure

When necessary to distinguish the dry pint or quart from the liquid pint or quart, the word "dry" should be used in combination with the name or abbreviation of the dry unit.

2 pints (pt)	= 1 quart (qt) = 67.2006 cubic inches
8 quarts	= 1 peck (pk) = 537.605 cubic inches = 16 pints
4 pecks	= 1 bushel (bu) = 2,150.42 cubic inches = 32 quarts

Avoirdupois Weight

When necessary to distinguish the avoirdupois ounce or pound from the troy ounce or pound, the word "avoirdupois" or the abbreviation "avdp" should be used in combination with the name or abbreviation of the avoirdupois unit.

(The "grain" is the same in avoirdupois and troy weight.)

$27\frac{11}{32}$ grains	= 1 dram (dr)
16 drams	= 1 ounce (oz) = $437\frac{1}{2}$ grains
16 ounces	= 1 pound (lb) = 256 drams = 7,000 grains
100 pounds	= 1 hundredweight (cwt)
20 hundredweights	= 1 ton = 2,000 pounds

In "gross" or "long" measure, the following values are recognized.

112 pounds	= 1 gross or long hundredweight
20 gross or long hundredweights	= 1 gross or long ton = 2,240 pounds

Tables of Equivalents

Lengths

1 angstrom (A)	0.1 nanometer (exactly)
	0.0001 micrometer (exactly)
	0.0000001 millimeter (exactly)
	0.000000004 inch
1 cable's length	120 fathoms (exactly)
	720 feet (exactly)
	219 meters
1 centimeter (cm)	0.3937 inch
1 chain (ch) (Gunter's or surveyor's)	66 feet (exactly)
	20.1168 meters
1 chain (engineer's)	100 feet
	30.48 meters (exactly)
1 decimeter (dm)	3.937 inches
1 degree (geographical)	364,566.929 feet
	69.047 miles (avg.)
	111.123 kilometers (avg.)
-of latitude	68.708 miles at equator
	69.403 miles at poles
-of longitude	69.171 miles at equator
1 dekameter (dam)	32.808 feet
1 fathom	6 feet (exactly)
	1.8288 meters (exactly)
1 foot (ft)	0.3048 meters (exactly)
1 furlong (fur)	10 chains (surveyors) (exactly)
	660 feet (exactly)
	⅛ statute mile (exactly)
	201.168 meters
1 inch (in)	2.54 centimeters (exactly)
1 kilometer (km)	0.621 mile
	3,281.5 feet

1 league (land)	3 survey miles (exactly)
	4.828 kilometers
1 link (Gunter's or surveyors')	7.92 inches (exactly)
	0.201 meter
1 link (engineers')	1 foot
	0.305 meter
1 meter (m)	39.37 inches
	1.094 yards
1 micrometer (μm)	0.001 millimeter (exactly)
[the Greek letter mu]	0.00003937 inch
1 mil	0.001 inch (exactly)
	0.0254 millimeter (exactly)
1 mile (mi) (statute or land)	5,280 feet (exactly)
	1,609 kilometers
1 international nautical mile (nmi)	1.852 kilometers (exactly)
	1.150779 survey miles
	6,076.11549 feet
1 millimeter (mm)	0.03937 inch
1 nanometer (nm)	0.001 micrometer (exactly)
	0.00000003937 inch
1 pica (typography)	12 points
1 point (typography)	0.013837 inch (exactly)
	0.351 millimeter
1 rod (rd), pole, or perch	16½ feet (exactly)
	5.029 meters
1 yard (yd)	0.9144 meter (exactly)

Areas or Surfaces

1 acre	43,560 square feet (exactly)
	4,840 square yards
	0.405 hectare
1 are (a)	119.599 square yards
	0.025 acre

1 bolt (cloth measure):	
length	100 yards (on modern looms)
width	42 inches (usually, for cotton)
	60 inches (usually, for wool)
1 hectare (ha)	2.471 acres
(1 square [building])	100 square feet
1 square centimeter (cm^2)	0.155 square inch
1 square decimeter (dm^2)	15.500 square inches
1 square foot (ft^2)	929.030 square centimeters
1 square inch (in^2)	6.4516 square centimeters (exactly)
1 square kilometer (km^2)	247.104 acres
	0.386 square mile
1 square meter (m^2)	1.196 square yards
	10.764 square feet
1 square mile (mi^2)	258.999 hectares
1 square millimeter (mm^2)	0.002 square inch
1 square rod (rd^2) sq. pole, or sq. perch	25.293 square meters
1 square yard (yd^2)	0.836 square meter

Capacities or Volumes

1 barrel (bbl) liquid	31 to 42 gallons
1 barrel (bbl), standard, for fruits,	7,056 cubic inches
vegetables, and other dry commodities	105 dry quarts
except dry cranberries	3.281 bushels, struck measure
1 barrel (bbl), standard, cranberry	5,826 cubic inches
	86 45/64 dry quarts
	2.709 bushels, struck measure
1 board foot (lumber measure)	a foot-square board 1 inch thick

1 bushel (bu) (U.S.) (struck measure)	2,150.42 cubic inches (exactly)
	35.239 liters
	2,747.715 cubic inches
(1 bushel, heaped [U.S.])	1.278 bushels, struck measure*
(1 bushel [bu] [British Imperial] [struck measure])	1.032 U.S. bushels struck measure
	2,219.36 cubic inches
1 cord (cd) firewood	128 cubic feet (exactly)
1 cubic centimeter (cm^3)	0.061 cubic inch
1 cubic decimeter (dm^3)	61.024 cubic inches
1 cubic inch (in^3)	0.554 fluid ounce
	4.433 fluid drams
	16.387 cubic centimeters
1 cubic foot (ft^3)	7.481 gallons
	28.317 cubic decimeters
1 cubic meter (m^3)	1.308 cubic yards
1 cubic yard (yd^3)	0.765 cubic meter
1 cup, measuring	8 fluid ounces (exactly)
	½ liquid pint (exactly)
1 dram, fluid (British)	0.961 U.S. fluid dram
	0.217 cubic inch
	3.552 milliliters
1 dekaliter (daL)	2.642 gallons
	1.135 pecks
1 gallon (gal) (U.S.)	231 cubic inches (exactly)
	3.785 liters
	0.833 British gallon
	128 U.S. fluid ounces (exactly)

*Frequently recognized as 1¼ bushels, struck measure.

(1 gallon [gal] British Imperial)	277.42 cubic inches
	1.201 U.S. gallons
	4.546 liters
	160 British fluid ounces (exactly)
1 gill (gi)	7.219 cubic inches
	4 fluid ounces (exactly)
	0.118 liter
1 hectoliter (hL)	26.418 gallons
	2.838 bushels
1 liter (L) (1 cubic decimeter exactly)	1.057 liquid quarts
	0.908 dry quart
	61.025 cubic inches
1 milliliter (mL) (1 cu cm exactly)	0.271 fluid dram
	16.231 minims
	0.061 cubic inch
1 ounce, liquid (U.S.)	1.805 cubic inches
	29.573 milliliters
	1.041 British fluid ounces
1 ounce, fluid (British)	0.961 U.S. fluid ounce
	1.734 cubic inches
	28.412 milliliters
1 peck (pk)	8.810 liters
1 pint (pt), dry	33.600 cubic inches
	0.551 liter
1 pint (pt), liquid	28.875 cubic inches (exactly)
	0.473 liter
1 quart (qt) dry (U.S.)	67.201 cubic inches
	1.101 liters
	0.969 British quart
1 quart (qt) liquid (U.S.)	57.75 cubic in (exactly)
	0.946 liter
	0.833 British quart

(1 quart [qt] [British])	69.354 cubic inches
	1.032 U.S. dry quarts
	1.201 U.S. liquid quarts
1 tablespoon	3 teaspoons (exactly)
	4 fluid drams
	$\frac{1}{2}$ fluid ounce (exactly)
1 teaspoon	$\frac{1}{3}$ tablespoon (exactly)
	$1\frac{1}{3}$ fluid drams

Weights or Masses

1 assay ton (AT)	29.167 grams
1 bale (cotton measure)	500 pounds in U.S.
	750 pounds in Egypt
1 carat (c)	200 milligrams (exactly)
	3.086 grains
1 dram avoirdupois (dr avdp)	
gamma, see microgram	$27\frac{11}{32}$ (= 27.344) grains
	1.772 grams
1 grain	64.799 milligrams
1 gram	15.432 grains
	0.035 ounce, avoirdupois
1 hundredweight, gross or long* (gross cwt)	112 pounds (exactly)
	50.802 kilograms
1 hundredweight, net or short (cwt. or net cwt.)	100 pounds (exactly)
	45.359 kilograms
1 kilogram (kg)	2.205 pounds
1 microgram (μg [The Greek letter mu in combination with the letter g])	0.000001 gram (exactly)
1 milligram (mg)	0.015 grain
1 ounce, avoirdupois (oz avdp)	437.5 grains (exactly)
	0.911 troy ounce
	28.350 grams

1 ounce, troy (oz t)	480 grains (exactly)
	1.097 avoirdupois ounces
	31.103 grams
1 pennyweight (dwt)	1.555 grams
1 pound, avoirdupois (lb avdp)	7,000 grains (exactly)
	1.215 troy pounds
	453.59237 grams (exactly
1 pound, troy (lb t)	5,760 grains (exactly)
	0.823 avoirdupois pound
	373.242 grams
1 ton, gross or long* (gross ton)	2,240 pounds (exactly)
	1.12 net tons (exactly)
	1.016 metric tons
1 ton, metric (t)	2,204.623 pounds
	0.984 gross ton
	1.102 net tons
1 ton, net or short (sh ton)	2,000 pounds (exactly)
	0.893 gross ton
	0.907 metric ton

*The gross or long ton and hundredweight are used commercially in the United States to only a limited extent, usually in restricted industrial fields. These units are the same as British "ton" and "hundredweight."

Notes

Part I: Essays on assays

Introduction

1. *Isaac Asimov's Book of Science and Nature Quotations* (London: Weidenfeld and Nicolson, 1988), p. 266.

Speed kills, but does sound?

1. Horace Parrack, "Effect of Air-borne Ultrasound on Humans," *International Audiology* 5 (1966), pp. 294-308.
2. Telephone interview with Dan Johnson, April 1, 1994.

How big is a big bang?

1. Morton Sternheim and Joseph Kane, *General Physics* (New York: John Wiley and Sons, 1986).

The metricsaurus

1. Ronald Zupko, *Revolution in Measurement* (Philadelphia: American Philosophic Society, 1990), pp. 20-25.
2. William Hallock, *Outline of the Evolution of Weights and Measures and the Metric System* (New York: The MacMillan Company, 1906).
3. Henri Moreau, *Le système métrique: des anciens mesures au système international d'unités* (Paris: Chiron, 1975), pp. 20-21.
4. Zupko, *Revolution in Measurement*, p. 139.
5. Moreau, *Le système métrique*, p. 29.
6. Denis Guedj, *La méridienne* (Paris: Seghers, 1987), p. 92.
7. Charles Treat, *A History of the Metric Controversy in the United States* (Washington, D.C.: U.S. Government Printing Office, 1971), p. 143.
8. Ibid., p. 83.
9. Ibid., p. 84.
10. *American Metric Journal* (1978), p. 285.
11. Treat, *A History of the Metric Controversy*, p. 143.
12. Maurice Dumesnils, *Etude critique du système métrique* (Paris: Gauthier-Villars, 1962), p.4.
13. J.W. Batchelder, *Metric Madness* (Old Greenwich, Conn.: Devin-Adair Co., 1981), p. 2.

The last metric artifact

1. W.E. Deeds, *Physics Today* (May 1993), pp. 90-91.

The earthquakesaurus

1. "Earthquake Magnitude Scales," *Nature*, February 8, 1990, p. 512.
2. Robert Temple, *The Genius of China* (New York: Simon & Schuster, 1986), p. 163.
3. Charles Richter, *Elementary Seismology* (San Francisco: W.H. Freeman & Co., 1958), p. 145.
4. *Engineering and Science* (March 1982), p. 27.

5. *Founders of Seismology*, p. 103.
6. *Engineering and Science* (March 1982), p. 28.
7. Charles F. Richter, "An Instrument Magnitude Scale," *Bulletin of the Seismological Society of America* 25 (1935), p. 27.
8. *Earthquakes and the Urban Environment*, p. 58.
9. *Engineering and Science* (March 1982), p. 25.
10. *Bulletin of the Seismological Society of America* (June 1992), p. 1306.

The cookingsaurus

1. Florence Moog, "Gulliver was a bad biologist," *Scientific American* (October 1949), pp. 52-55.
2. Betty Crocker, *Betty Crocker's Picture Cookbook* (New York: McGraw-Hill and General Mills Inc., 1950), p. 6.
3. Katherine Grover, ed., *Dining in America*, 1850-1900 (Amherst: University of Massachusetts Press, 1987), p. 112.
4. Elizabeth Raper, *Receipt Book of Elizabeth Raper* (London: Nonesuch Press, 1924), p. 41.
5. Eric Quayle, *Old Cook Books* (London: Studio Vista, 1978), p. 155.
6. Jane Cooper, *Woodstove Cookery* (Charlotte, VT: Garden Way, 1977), p. 71.
7. Mrs. Beeton, *Mrs. Beeton's Cookery and Household Management* (London: Ward Lock, 1980, originally published 1861), p. 107.
8. E. Neige Todhunter, "Seven Centuries of Cookbooks," *Nutrition Today*, January/February 1992, p. 10.
9. Fannie Farmer, ed., *The Boston Cooking-School Cook Book* (New York: New American Library Books, 1974, originally published 1896), p. 1.
10. Ibid., p. 1.
11. For biographical information on Fannie Farmer, see Laura Shapiro's *Perfection Salad* (New York: Farrar, Strauss & Giroux, 1986) and Russel Lynes's chapter on her in *The American Heritage Cookbook* (New York: American Heritage Publishing, 1964).
12. Lynes, *American Heritage Cookbook*, p. 297.
13. M.V. Tracy, "Food for Thought," *Perspectives in Biology and Medicine*, Autumn 1989, p. 54.
14. Ibid., p. 55.
15. Harvey Levenstein, *Revolution at the Table* (New York: Oxford University Press, 1988), p. 56.
16. John Hess and Karen Hess, *The Taste of America* (New York: Viking, 1977), pp. 113-14.
17. "Housekeepers Vote for Standard Measures," *Good Housekeeping*, September 1923, p. 80.
18. Minna Denton, "The Why of the Oven Thermometer," *Ladies' Home Journal*, July 1922, pp. 86-87.
19. Ethel Dickinson, "Brimful, Not Running Over," *Good Housekeeping*, June 1923, p. 72.

20. Nella Whitfield, *Kitchen Encyclopedia* (London: Spring Books, n.d.), p. 148.
21. Rupert Croft-Cooke, *Cooking for Pleasure* (London: Collins, 1963), p. 9.
22. B. Haughton et al., "A Historical Study of the Underlying Assumptions for United States Food Guides from 1917 through the Basic Four Food Group Guide," *Journal of Nutrition Education*, July/August 1987, p. 170.
23. Ibid., p. 173.
24. Kenneth Carpenter, "The History of Enthusiasm for Protein," *Journal of Nutrition*, 116 (1986), p. 1364.
25. Ann Walker, ed., *Applied Human Nutrition* (New York: Ellis Horwood, 1990), p. 5.

The ozonesaurus

1. Manfred Tevini, ed., *UV-B Radiation and Ozone Depletion* (Boca Raton: Lewis Publishers, 1993), p. 99.
2. Ibid., pp. 40-42.

Part II: The sizesaurus

The quick-and-dirty wordsaurus dictionary

1. Morton Sternheim and Joseph Kane, *General Physics* (New York: John Wiley & Sons Inc., 1986), p. 192.
2. Paul Hewitt, *Conceptual Physics*, 5th edition (Boston: Little, Brown, 1905) pp. 349, 352.
3. *The Guinness Book of Records* (New York: Bantam, 1990), p. 74.
4. E.V.P. Smith and K.C. Jacobs, *Introductory Astronomy and Astrophysics* (Philadelphia: Saunders College Publishing, 1973), p. 528. See also R.C. Weast and D.R. Lide, eds., *The CRC [Chemical Rubber Company] Handbook of Chemistry and Physics*, 70th edition (West Palm Beach, FL: CRC Press, 1990), p. F-182.
5. H. Arthur Klein, *The World of Measurements* (New York: Simon & Schuster, 1974), p. 197.
6. Tom Parker, *Rules of Thumb* (Boston: Houghton-Mifflin, 1983), rule no. 196.
7. Dan Schroeder, Weber State College, personal communication.
8. Environment Canada, *Primer on Ozone Depletion* (Ottawa: Environment Canada, 1993), pp. 11-12.
9. Vaclav Smil, *General Energetics* (New York: John Wiley, 1991), p. 112.
10. Sternheim and Kane, *General Physics*, p. 211.
11. Sternheim and Kane, *General Physics*, p. 58.
12. Weast and Lide, *CRC Handbook*, p. E-45.
13. Edward Purcell, "Back of the Envelope," *American Journal of Physics*, April 1984, p. 301.
14. S. Singh and T. Voss, "Drop Heights Encountered in the United Parcel Service Small Parcel Environment in the United States," *Journal of Testing and Evaluation*, September 1992, pp. 382-87.

15. David Bodnais, *The Secret House* (New York: Simon & Schuster, 1986), p. 131.
16. Donald Proctor and Ib Andersen, eds., *The Nose* (New York: Elsevier, 1982), p. 320.
17. *The New York Public Library Desk Reference* (New York: Simon & Schuster, 1989), p. 131.
18. Thomas Rossing, *The Science of Sound* (Reading, MA: Addison-Wesley, 1990), p. 179.
19. Klein, *The World of Measurements*, p. 215.
20. Sternheim and Kane, *General Physics*, p. 547.
21. David Halliday and Robert Resnick, *Fundamentals of Physics* (New York: John Wiley, 1981), p. 166.
22. Albert Bartlett, "Etcetera," *The Physics Teacher*, April 1987, p. 256.
23. *The Guinness Book of Records*, p. 52.
24. René Racine, University of Montreal, personal communication.
25. *The Guinness Book of Records*, p. 77.
26. James Trefil, *Sharks Have No Bones* (New York: Simon & Schuster, 1992), pp. 133-34.

A measuring macropedia

1. Clark Chapman, "Hazard to Civilization of Asteroid and Cometary Impacts," paper presented at a conference on asteroid impacts, St. Petersburg, Florida, October 1990, p. 9.
2. Horace Parrack, "Effect of Air-borne Ultrasound on Humans," *International Audiology* 5 (1966), pp. 294-308.

Conclusion: A silly end is better than no end

1. From *The Journal of Irreproducible Results* (1965), cited in Philip Simpson, *A Random Walk Through Science* (New York: Crane, Russal and Company, 1973).
2. Amos Shapir, Department of Computer Science, The Hebrew University of Jerusalem, personal communication.
3. From *NBS Standard* 1 (January 1970), cited in Simpson, *A Random Walk Through Science.*
4. *Harper's*, May 1984, p. 48.
5. Ibid., p. 47.

Bibliography

Achenbach, Joe. *Why Things Are.* New York: Ballantine Books, 1991.
Becker, Robert. *Cross Currents.* Los Angeles: Jeremy Tarcher, 1993.
Becker, Robert and Andrew Marino. *Electromagnetism and Life.* Albany, N.Y.: State University of New York Press, 1982.
Berg, Paul and Maxine Singer. *Dealing with Genes.* Mill Valley, California: University Science Books, 1992.
Bodanis, David. *The Secret House.* New York: Simon and Schuster, 1986.
Brancazio, Peter. *Sport Science.* New York: Simon and Schuster, 1984.
Brown, John. Talk at the American Association for the Advancement of Science, Boston, 1993.
Brown, Lester. *State of the World 1991.* New York: W.W. Norton, 1991.
Carlisle, Jon. *In the World of Science.* Happague, N.Y.: Barron's Educational Series, 1992.
Castonguay, Rino and Leonard Gallant. *$E = MC^2$: Introduction à la physique.* Montreal: Editions du Renouveau Pedagogique, 1991.
Charron, David. *Humidex and Its Use in the Workplace.* Hamilton, Ont.: Canadian Centre for Occupational Health and Safety, 1989.
Cheremisinoff, Nicholas. *Industrial Hazards Desk Book.* Lancaster, Penn.: Technomic Publishing Company, 1987.
Clader, William. *Sex, Function and Life History.* Cambridge, Mass.: Harvard University Press, 1984.
Davidovits, Paul. *Physics in Biology and Medicine.* New York: Prentice Hall, 1975.
The Diagram Group. *Comparisons.* New York: St. Martins Press, 1980.
Dorf, Richard. *The Energy Fact Book.* New York: McGraw-Hill Book Company, 1980.
Encyclopedia Britannica, 15th edition. Chicago: Encyclopedia Britannica Inc., 1989.
Encyclopedia of Physical Science and Technology, vol. 1. Orlando: Academic Press, 1987.
Faissler, William. *An Introduction to Modern Electronics.* New York: John Wiley & Sons, Inc., 1991.
Famighetti, Robert, ed. *The World Almanac and Book of Facts 1994.* Mahwah, N.J.: Funk & Wagnalls Corporation, 1994.
Fierer, John. *SI Metric Handbook.* New York: Charles Scribner and Sons and Charles Bennet Co., 1977.
Fox, Brian and Allan Cameron. *Food Science: A Chemical Approach.* London: Hodder and Stoughton, 1982.
Glover, Thomas. *DeskRef.* Littleton, Colorado: Sequoia Publishing, 1993.
Gold, Carol, ed. *Foodworks.* Toronto: Kids Can Press, 1986.
———. *How Sport Works.* Toronto: Kids Can Press, 1988.
Gordon, J.E. *Structures, Or Why Things Don't Fall Down.* London: Penguin, 1978.
Guinness Book of Records. Enfield, England: Guinness Publishing, 1987, 1990, 1993, and 1994.
Guinness Book of Weather Facts and Feats, 2nd edition. Enfield, England: Guinness Publishing, 1982.
Halliday, David and Robert Resnick. *Fundamentals of Physics.* New York: John Wiley and Sons, 1988.

The Handy Science Answer Book. Compiled by the Science and Technology Department of the Carnegie Library of Pittsburgh. Visible Ink Press, 1994.

Hewitt, Paul. *Conceptual Physics*, 5th edition. Boston: Little, Brown and Company, 1985.

Hirsch, Alan. *Physics: A Practical Approach.* Toronto: John Wiley and Sons, 1991.

Houwink, R. *The Odd Book of Data.* London: Elsevier, 1965.

IES Lighting Handbook. New York: Illuminating Engineering Society of North America, 1981.

Ingram, Jay. *The Science of Everyday Life.* Markham, Ont.: Penguin, 1989.

Jerrard, H.G. and D.B. McNeill, eds. *The Dictionary of Scientific Units.* New York: Chapman and Hall, 1980.

Johnston, William. *For Good Measure.* New York: Holt Rinehart, 1975.

Journalists' Guide to Nuclear Power. Toronto: Ontario Hydro, 1988.

Kavalar, Lucy. *Freezing Point: Cold As a Matter of Life and Death.* New York: John Day & Company, 1972.

Keating, Michael. *Covering the Environment.* London, Ontario: University of Western Ontario, 1992.

Kelves, Daniel and Leroy Hood. *The Code of Codes.* Cambridge, Mass.: Harvard University Press, 1993.

Klein, H. Arthur. *The World of Measurements.* New York: Simon and Schuster, 1974.

Krantz, Les. *What The Odds Are.* New York: HarperCollins, 1992.

Laing, D.G., et al. *The Human Sense of Smell.* Berlin: Springer-Verlag, 1991.

Lambert, Barrie. *How Safe Is Safe?* London: Unwin Hyman Ltd., 1990.

Leckie, Jim, Gil Master, Harry Whitehouse and Lily Young. *More Other Homes and Garbage.* San Francisco: Sierra Club, 1981.

Marion, Jerry. *Energy in Perspective.* New York: Academic Press, 1974.

McCutcheon, Marc. *The Compass in Your Nose.* Los Angeles: Jeremy Tarcher, Inc., 1989.

Mohl, Norman et al. *A Textbook of Occlusion.* Chicago: Quintessence Publishing Co., 1988.

Moncrieff, R.W. *Odours.* London: William Heinemann Medical Books Ltd., 1970.

National Research Council press release. Ottawa: Mar. 8, 1993.

New York Library Desk Reference. New York: Simon and Schuster, 1989.

Nobile, Philip and John Deedy, eds. *The Complete Ecology Fact Book.* Garden City, N.Y.: Doubleday & Co., 1972.

Ohanian, Hans. *Physics.* New York: W.W. Norton, 1985.

Parker, James and Vita West, eds. *Bioastronautics Data Book.* Washington, D.C.: NASA Publications, 1973.

Parker, Tom. *Rules of Thumb.* New York: Houghton Mifflin, 1983.

The Physics Teacher (various issues). College Park, Md.: The American Association of Physics Teachers.

Primer on Ozone Depletion. Ottawa: Environment Canada, 1993.

Proctor, Donald and Ib Andersen. *The Nose.* London: Elsevier, 1982.

Rossing, Thomas. *The Science of Sound.* Reading, Mass.: Addison-Wesley Publishing Company, 1990.

Scarne, John. *Scarne on Cards.* New York: Signet Books, 1965.
Schmidt-Nielsen, Knut. *Scaling: Why is Animal Size So Important.* Cambridge: Cambridge University Press, 1984.
Serway, Raymond. *Physics for Scientists and Engineers.* Philadelphia: Saunders College Publishing, 1990.
Shapiro, Robert. *The Human Blueprint.* New York: St. Martins Press, 1991.
Sheftel, Richard, ed. *ABC's of Nature.* Montreal: Reader's Digest, 1984.
Simpson, Philip. *A Random Walk Through Science.* New York: Crane, Russal and Company, 1973.
Sliwa, Jan and Leslie Fairweather, eds. *AJ Metric Handbook.* London: The Architectural Press, 1973.
Smil, Vaclav. *General Energetics.* New York: John Wiley & Sons, 1991.
Smith, E.V.P. and K.C. Jacobs. *Introductory Astronomy & Astrophysics.* Philadelphia: W.B. Saunders Company, 1973.
Smith, Norman. *Energy Isn't Easy.* New York: Coward-McCann, 1984.
Snyder, Carl. *The Extraordinary Chemistry of Ordinary Things.* New York: John Wiley and Sons, 1992.
Sternheim, Morton and Joseph Kane. *General Physics.* New York: John Wiley & Sons Inc., 1986.
Stoddart, M. Michael. *The Ecology of Vertebrate Olfaction.* New York: Chapman and Hall, 1990.
Sumner, David, Tom Wheldon and Walter Watson. *Radiation Risks: An Evaluation.* Glasgow: Tarragon Press, 1991.
Trefil, James. *Sharks Have No Bones.* New York: Simon and Schuster, 1992.
Tributsch, Helmut. *How Life Learned to Live.* Cambridge, Mass.: The MIT Press, 1982.
van Bergeijk, Willem, John Pierce and David Edward. *Waves and the Ear.* Garden City, New York: Anchor Books, 1960.
Voluntary Product Standards PS-73-89, Glass Bottles for Carbonated Soft Drinks. Washington, D.C.: U.S. Department of Commerce, 1989.
Vincent, Julian. *Structural Biomaterials.* Princeton, N.J.: Princeton University Press, 1990.
Vogel, Steven. *Life's Devices.* Princeton, N.J.: Princeton University Press, 1988.
———. *Vital Circuits.* Oxford: Oxford University Press, 1992.
Voyager Neptune Travel Guide. Washington, D.C.: National Aeronautics and Space Administration, 1989.
Walker, Jearl. *The Flying Circus of Physics.* New York: John Wiley & Sons, 1977.
Weast, R.C. and P.R. Lide, eds. *CRC Handbook of Chemistry and Physics,* 70th edition. Florida: CRC Press, 1990.
Weaver, Warren. *Lady Luck.* New York: Anchor Books, 1963.
Wils, Christopher Exons. *Introns and Talking Genes.* New York: Basic Books, 1991.
Wright, R.H. *The Sense of Smell.* Boca Raton: CRC Press, 1982.
Wurman, Richard Saul. *Information Anxiety.* New York: Bantam, 1990.

Index